CW0641908

Smart Futures, Challenges of Urbanisation, and Social Sustainability

Mohammad Dastbaz · Wim Naudé
Jamileh Manoochehri

Editors

Smart Futures, Challenges of Urbanisation, and Social Sustainability

 Springer

Editor
Mohammad Dastbaz
University of Suffolk
Ipswich, UK

Jamileh Manoochehri
Leicester School of Architecture
De Montfort University Leicester
School of Architecture
Leicester, UK

Wim Naudé
Maastricht School of Management
and Maastricht University
Maastricht, The Netherlands

IZA Institute of Labor Economics
Bonn, Germany

ISBN 978-3-319-74548-0   ISBN 978-3-319-74549-7   (eBook)
https://doi.org/10.1007/978-3-319-74549-7

Library of Congress Control Number: 2018935403

Printed on acid-free paper

This Springer imprint is published by the registered company Springer International Publishing AG, part of Springer Nature.
The registered company address is: Gewerbestrasse 11, 6330 Cham, Switzerland

Preface

The idea behind putting this volume together was to provide a considered approach in dealing with some of the most pressing issues facing the development of our societies. We believe that social sustainability is one of the least understood of the different areas of sustainable development. Large migrations of populations from rural areas to cities without planning and without appropriate infrastructure, including housing, education, and health care requirements, have created significant challenges across the globe. There is now the need to look at how the intelligent use of data on population growth and movement in planning, design, and constructing infrastructure can be used to meet the challenges of urbanization and the sustainable development of societies in the face of rapid population growth.

Unlike the political games being played at global stage, the science does not believe that the world faces a "fake challenge" and does not deny the damage we have done to our environment. Furthermore, sustainable development is not just about climate control and more about a measured approach to addressing the 17 "Sustainable Development Goals" identified by the United Nations (announced as part of "the 2030 Agenda for Sustainable Development" adopted at the United Nations Sustainable Development Summit on 25 September 2015).

The issues we face are extremely serious and require collective action to avert a global catastrophe. The Paris climate agreement and its ratification was only the first step in the long road to addressing the short-, medium-, and long-term issues around our sustainable development. Without proper monitoring and holding governments to account for their actions and targets, the damage to our sustainable environment and future will continue and we will all have our share or burden and blame if we do not act now.

Ipswich, UK   Mohammad Dastbaz

Contents

Part I
Urbanisations and Social Sustainability

Chapter 1
A 2030 Vision: "Fake Challenge" or Time for Action

Mohammad Dastbaz

"When the last tree is cut, the last fish is caught, and the last river is polluted; when to breathe the air is sickening, you will realize, too late, that wealth is not in bank accounts and that you can't eat money…" – Alanis Obomsawin "an Abenaki from the Odanak reserve" – From the collection of essays published in 1972 titled "Who is the Chairman of This Meeting?"

1.1 Context

In June 2017, Donald Trump announced that the USA will be withdrawing from the Paris climate agreement, to put "America first" and to save "American jobs", calling the agreement unfair to America's interest. At the time, the US government, Syria and Nicaragua stood as the only other countries who had not signed the accord (Syria has since joined the agreement). This meant that the USA, the world's second largest emitter of greenhouse gases, moved to jeopardise the Paris Agreement, effectively denying that there is an environmental challenge that needs to be dealt with and saw no need and felt no responsibility to help the world and humanity address our ever-growing challenge of climate change and environmental damage.

The Paris Agreement was the result of long and hard negotiations that were done before and during the 2015 "United Nations Climate Change Conference, COP 21" which was held in Paris, between 30 November and 12 December 2015. Following the 1997 Kyoto Protocol, this was to be the most important initiative by countries of the world to address the ever-growing problems of global warming and environmental challenges facing future and sustainable development of our planet.

M. Dastbaz (✉)
University of Suffolk, Ipswich, UK
e-mail: m.dastbaz@uos.ac.uk

© Springer International Publishing AG, part of Springer Nature 2018
M. Dastbaz et al. (eds.), *Smart Futures, Challenges of Urbanisation, and Social Sustainability*, https://doi.org/10.1007/978-3-319-74549-7_1

The conference's negotiated outcome was an agreement to address climate change, representing a consensus between 196 countries and organisations attending it. The Paris Agreement was later, on 22 April 2016 (Earth Day), signed by 174 countries in New York.

The "United Nations Framework Convention on Climate Change", web portal in 2016, reported that of the 197 Parties to the Paris Convention, 169 Parties have ratified the agreement, and therefore, on 5 October 2016, the threshold for entry into force of the Paris Agreement was achieved, and the Paris Agreement entered into force on 4th November 2016.[1]

The Paris Agreement recognised the need "for an effective and progressive response to the urgent threat of climate change on the basis of the best available scientific knowledge; Also recognizing the specific needs and special circumstances of developing country Parties, especially those that are particularly vulnerable to the adverse effects of climate change, as provided for in the Convention…".[2]

More importantly the signatories to the Paris Agreement acknowledge that "Climate change is a common concern of humankind, Parties should, when taking action to address climate change, respect, promote and consider their respective obligations on human rights, the right to health, the rights of indigenous peoples, local communities, migrants, children, persons with disabilities and people in vulnerable situations and the right to development, as well as gender equality, empowerment of women and intergenerational equity…".[3]

1.2   Social Sustainability and Urbanisation

There has been significant body of research around what the literature calls "urbanisation" and the challenges this poses towards our sustainable development. It is generally viewed that sustainable development and the ever-growing population movement from rural areas to cities pose one of the key challenges in setting goals and putting plans in place to deal with the twenty-first century changing population landscape and dwindling natural resources.

Dempsey et al. (2009) state that "Sustainable development is a widely used term, which has been increasingly influential on UK planning, housing and urban policy in recent years. Debates about sustainability no longer consider sustainability solely as an environmental concern, but also incorporate economic and social dimensions".[4]

---

[1] http://unfccc.int/paris_agreement/items/9485.php

[2] http://unfccc.int/files/essential_background/convention/application/pdf/english_paris_agreement.pdf

[3] http://unfccc.int/files/essential_background/convention/application/pdf/english_paris_agreement.pdf

[4] The social dimension of sustainable development: Defining urban social sustainability – http://onlinelibrary.wiley.com/doi/10.1002/sd.417/full

A report by the United Nations titled "World Urbanisation Prospects" published in 2014[5] warns against the rapid urbanisation of our society predicting what while in the 1950s 30% of the population lived in the cities, this grew to 54% in 2014, and it is predicted that by 2050, 66% of the population will be living in urban areas. There are also a marked difference and interesting emerging urbanisation trends between developing and developed countries. While in North America and Europe, 82% and 73%, respectively, of their population live in urban area, Asia and Africa remain mainly rural with 40% and 48% of their population living in rural area. The report further observes that both Africa and Asia are seeing the fastest urbanisation trends than any other continent, predicting that as much as 64% of their population could be living in urban areas by 2050.

It is therefore not surprising that as more and more of the population moves from the rural to urban areas, due to changes in economic wealth creation, the technological revolution has significantly transformed the way we live, produce and distribute products and manage our lives, and without proper long-term planning and having the required infrastructure including water, electricity, housing, healthcare and education in place, serious new challenges are emerging that are increasingly difficult to manage.

The United Nations "Habitat III" conference, on "Housing and Sustainable Urban Development", held between the 17th and 20th of October 2016 in Quito (the capital city of Ecuador), brought together over 30,000 people from 167 countries including more than 2000 representatives of local and regional governments. The conference noted that "by 2050, the world's urban population is expected to nearly double, making urbanization one of the twenty-first century's most transformative trends. Populations, economic activities, social and cultural interactions, as well as environmental and humanitarian impacts, are increasingly concentrated in cities, and this poses massive sustainability challenges in terms of housing, infrastructure, basic services, food security, health, education, decent jobs, safety and natural resources, among others".[6]

As Fig. 1.1 shows, one of the key aspects of growing urbanisation is the transformation of human settlement arrangements and the creation of more and more mega and large cities.

1.3 Millennium Agenda and Sustainable Development Goals

The "Millennium Summit" of the United Nations in 2000 established the "Millennium Development Goals" (MDGs) that set the agenda for developing the sustainability debate across the world. Eight MDGs were identified listed below:

---

[5] https://esa.un.org/unpd/wup/Publications/Files/WUP2014-Highlights.pdf

[6] http://habitat3.org/wp-content/uploads/NUA-English.pdf

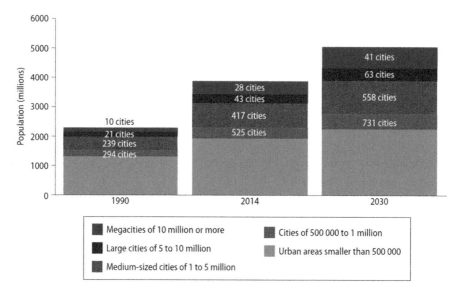

Fig. 1.1 Global urban population and associated growth of cities of all sizes (Source: https://esa.un.org/unpd/wup/Publications/Files/WUP2014-Highlights.pdf)

1. Eradicate poverty and hunger (MDG1)
2. Achieve universal primary education (MDG2)
3. Gender equality and empowering women (MDG3)
4. Reduce child mortality (MDG4)
5. Improve maternal health (MDG5)
6. Combat HIV/AIDS, malaria and other diseases (MDG6)
7. Environmental sustainability (MDG7)
8. Global partnership and development (MDG8)[7]

Furthermore, the UN General Assembly, 25 September 2015, adopted a resolution titled "Transforming our world: the 2030 Agenda for Sustainable Development", which identified significant challenges to our future sustainable development over the next 15 years. These included extreme poverty, as one of the greatest nemesis humanity face in the twenty-first century. Seventeen Sustainable Development Goals (SDGs) – Fig. 1.2 – and 169 targets were identified by the UN General Assembly.[8]

The key issues covered in this book relate to MDG7 (environmental sustainability) and SDGs 3, 4, 6, 8, 11 and 17. The chapters in this book are divided into two board sections. Section 1.1 deals with conceptual issues related to the above SDGs and section provides a number of relevant case studies.

---

[7] Dastbaz et al. (2016). Building sustainable futures design and the built environment. Springer Publications.

[8] In Dastbaz et al. (2017). Building information modelling, building performance, design and small construction. Springer publication.

Fig. 1.2 UN Sustainable Development Goals (Source: http://www.un.org/sustainabledevelopment/sustainable-development-goals/)

1.4   Planning, Policy and Practice, Urbanisation and Alienation

Ian Strange in his chapter "Urban Planning and City Futures: Planning for Cities in the Twenty-First Century" argues that historically, cities have been the powerhouses of economic growth and development, the centres of major social and cultural change, and the sites of rapid movement and change in population. He further elaborates that a key question to ask is: How do cities, and those who plan for them, respond to these challenges so that urban futures produce fairer and more just places? Equally important is to ask: What are the technological challenges that are faced by urban planners and decision-makers as they search for solutions to complex and multi-faceted urban problems while cities transition from their analogue pasts to their digital futures? What forms of governance and local democracy allow greater degrees of citizen involvement and participation in the making of urban futures? And, what is the role of the state in shaping the policy and political context within which cities can plan for their futures? Addressing these questions is the focus of Prof. Strange's chapter.

Chapter 3 deals with the process of development, urbanisation and entrepreneurship. Wim Naudé argues that both urbanisation and entrepreneurship are central "demographic facts" of the early twenty-first century. No economy has developed without urbanising. Furthermore Prof. Naudé notes that entrepreneurs hasten urbanisation, but urbanisation also benefits entrepreneurs, so that a virtuous cycle comes into being. The ensuing process of structural transformation is one of the salient facts of development, wherein entrepreneurs create jobs in urban areas, in new sectors such as industry or services, attracting labour and other entrepreneurs out of the agricultural sector and out of rural areas. Eventually, if these developing urban areas and their entrepreneurial dynamics can be appropriately managed and incentivised, it may lead to improvements in GDP per capita, and many other improvements in the quality of life.

The issue of social sustainability, housing and alienation is treated in Chap. 4, where Dr. Manoochehri argues the relevance of the sustainability agenda at a time when post-industrial capitalism dominates economic and social relations. It furthermore observes that attempts to find links between the way housing is understood as a need and a right, while it is, at the same time, defined increasingly as a commodity then the by-product of this condition could result in "alienation". The loss of access to adequate housing of socially acceptable standards and the dispersal of people from communities are turning into a new norm in the twenty-first century UK. The chapter proposes that this relates most closely with the commodification and marketisation of housing and changing significance from its use value to exchange value.

Chapter 5, by Michael Crilly and Professor Mark Lemon, examines the misalignment of policy and practice in sustainable urban design. It states that a so-called urban renaissance throughout the UK's major cities began in the late 1990s as a policy response to the dominant trends in counter-urbanisation and inner-city decline. According to the authors, this urban policy encompassed the interconnected themes of sustainability and design quality as the basis for public sector investments into housing and urban infrastructure. The authors explore these ideas through a chronology of urban regeneration and neighbourhood planning policies, and in so doing, it tracks the progression of sustainability and design quality and community engagement in the planning system and their involvement in urban redevelopment and supporting policy. These themes are presented from the perspective of an urban design practitioner who has spent much of the past 20 years working through the changing policy landscape.

The challenges of "Reporting corporate sustainability" and "the organisational and political rhetoric" are treated comprehensively by Profs. Gorse and Sturges and Dr. Duwebi and Mike Bates in Chap. 6. The authors argue that political rhetoric and regulatory change brings uncertainty to those industrial organisations that have aligned their strategies to take advantage of sustainable practice. These developments are too recent to predict the impact on the industry and corporate sustainability; nevertheless, the position that industry has adopted prior to these changes is interesting. Will those organisations already committed to sustainability continue maintaining a social or corporate interest or will the changes bring sustainable and economic uncertainty?

1.5   Spatial Planning, Smart Eco-Cities, Agritech, Safety Online and Sustainable Education

Chapter 7 in Sect. 7.2 of the book provides a fascinating case study about social-environmental interface of sustainable development in Libya. Dr. Taki and Jamal Alabid argue that one of the key challenges in the spatial planning and design of urban settlements in desert housing is the trade-off between the harshness of the environment and meeting social needs concurrently. The chapter further discusses the role of socio-environmental dimensions on planning and developing urban forms within the context of the Sahara Desert. The case study of Ghadames old settlements integrates spatial planning methodology with environmental thoughts, in accordance with sociocultural values examined in this chapter.

Prof. Cilliers presents an interesting case study on "The undervaluation, but extreme importance, of social sustainability in South Africa". The author observes that almost every planning-related paper employs the increasing urbanisation figures and problematic impact of such, to substantiate research needs and approaches. This is especially true for the African continent which often tops the charts in terms of population growth. Accordingly, spatial planning recently became a tool to guide broader sustainability thinking and direct the planning of smart futures. Within this notion, spatial planning, theoretically, often relates to the three dimensions of sustainability (economic, environmental and social) to find adequate planning solutions. In practice divergent scenarios are seen, where focus is often placed on economic and environmental interventions, as easier implementable solutions, in comparison to more complex approaches related to social sustainability.

Chapter 9, by Dr. van Dijk, provides a case study on "Smart eco-cities are managing information flows in an integrated way, the example of water, electricity and solid waste". The author observes that smart eco-cities are about managing flows of information in an integrated way. The information may concern the traffic, the people, pollution or the number of enterprises moving in and out of the city. We focus on a classification of cities. Resilient cities are a more defensive concept, while the eco city concept translates an ambition what a city should be. We will argue that the smart eco city concept integrates the two approaches. The chapter further provides the examples of water, electricity and solid waste management that will be used to show how cities can be smarter by using the available information differently.

Helen Mytton-Mills, from Aponic, UK, provides a very interesting "Agritech case study of Aeroponics". The author points out that global resources for food production are heavily, and unsustainably, utilised with known inefficiencies, in a complex, integrated international supply chain. Product availability and scarcity vary from country to country and continent to continent, approximately correlated with GDP rather than with either population quanta or domestic agricultural resources. The author further points out that the challenges in evening up global nutrition standards with the additional expected demand require great change within the agriculture industry, especially if the outputs are to be produced sustainably in order to safeguard future generations of production and ecosystem health. The author goes on to argue that although great efforts have been taken to make more of less, however, a complete rethink to resource usage is key to achieving the sustainability required alongside raising food standards globally.

Simon Dukes, CEO of Cifas, provides an insightful chapter on "Safety and Cyber Security in a Digital Age". He observes that in October 2014, at an event in Canary Wharf in London, Martin Smith MBE, chairman of the protective security consultancy, said the following: "if I throw a brick through a jewellery shop window and steal a watch, it is not a 'brick-crime.' If we insist on calling significant and increasing instances of fraud, financial scams and illegal pornography, 'cyber-crimes' we risk obfuscation by focusing the eye not on the crime but on the manner of its conveyance". The chapter deals in depth with growing challenges of "cyber-crime" and "identity theft" in the twenty-first century and the challenges that this provides for societies.

Chapter 13 examines the "A sustainable higher education sector: the place for mature and part-time students?". Dr. Richardson notes that in England, explicit policy efforts to improve the access of those groups under-represented in higher education (HE) have formed a key element of national education policies for over fifty years. It further observes that it is estimated that 80% of new jobs will require degree level qualifications by 2020, a target which cannot be met solely by those 40% of young people who access HE. This posits the question of how such economic goals are to be met without the reinvigoration of access for part-time mature learners.

The final chapter of this book provides a fascinating insight into "Buildings that perform: thermal performance and comfort". Prof. Gorse et al. note that developers, designers and contractors are increasingly using titles such as "green", "eco" and "low energy" to describe their buildings and reassure the environmentally conscious consumer of the green credentials of the property that they are investing in. Unfortunately, relatively few construction companies engage in research and development (R&D) to underpin their marketing rhetoric. The chapter reports on the tests and monitoring undertaken on buildings and the common issues that they are designed to interrogate and presents cases where developers are taking measures to achieve homes that function as expected and meet the requirements of the occupants, thus fulfilling their "green" credentials.

1.6   Final Note

The idea behind putting this volume together was to provide a considered approach in dealing with some of the most pressing issues facing the development of our societies. Unlike the political games being played at global stage, the science does not believe that the world faces a "fake challenge" and does not deny the damage we have done to our environment. The issues we face are extremely serious and require collective action to avert a global catastrophe. The Paris climate agreement and its ratification were only the first step in the long road to addressing the short-, medium- and long-term issues around our sustainable development. Without proper monitoring and holding governments to account for their actions and targets, the damage to our sustainable environment and future will continue, and we will all have our share or burden and blame if we do not act now.

References

Dastbaz, M., Gorse, C. (2016); "Sustainable Ecological Engineering Design: Selected Proceedings from the International Conference of Sustainable Ecological Engineering Design for Society (SEEDS)" – Springer Publications, 2016

Dastbaz, M., Muncaster, A. Gorse, C. (2017); "Building Information Modelling, Building Performance, Design and Smart Construction" Edited book, Springer Publications, 2017

Dempsey, N., Bramley, G., Power, S., Brown, C., (2009); "The Social Dimension of Sustainable Development: Defining Urban Social Sustainability". Sustainable Development doi: 10.1002/sd.417. Department for Communities And Local Government (DCLG), 2006a. Planning Obligations: Practice Guidance, DCLG, London.

Chapter 2
Urban Planning and City Futures: Planning for Cities in the Twenty-First Century

Ian Strange

2.1 Introduction

Historically, cities have been the powerhouses of economic growth and development, the centres of major social and cultural change, and the sites of rapid movement and change in population. As we move more and more into the twenty-first-century cities will continue to be the focus of economic activity, the key spaces of social and cultural interaction, and the home for larger and larger urban populations. The first half of the twenty-first century is throwing up major challenges to cities, to their populations, and to those who seek to plan for their development. These challenges include a rapid and rampant process of economic growth and restructuring that often leaves some places as 'winners' and others as 'losers', the continuing pressures of urbanisation and demographic change, the call for the development of sustainable forms of urban transportation and infrastructure, the need to provide more secure and affordable homes, and the rising tide of local accountability as communities seek greater involvement and participation in local decision-making. A key question to ask is: How do cities, and those who plan for them, respond to these challenges so that urban futures produce fairer and more just places? Equally important is to ask: What are the technological challenges that face urban planners and decision-makers as they search for solutions to complex and multi-faceted urban problems while cities transition from their analogue pasts to their digital futures? What forms of governance and local democracy allow greater degrees of citizen involvement and participation in the making of urban futures? And, what is the role of the state in shaping the policy and political context within which cities can plan for their futures? Addressing these questions will be the focus of this chapter. Using examples from cities in the UK and elsewhere, the chapter will explore and analyse the complexities and difficulties of planning in, and for, the twenty-first-century city.

I. Strange (✉)
School of the Built Environment & Engineering, Leeds Beckett University, Leeds, UK
e-mail: i.strange@leedsbeckett.ac.uk

© Springer International Publishing AG, part of Springer Nature 2018
M. Dastbaz et al. (eds.), *Smart Futures, Challenges of Urbanisation, and Social Sustainability*, https://doi.org/10.1007/978-3-319-74549-7_2

13

2.2   Global Urbanisation and Development of Cities

It is a truism that we live in a global and increasingly urbanised world. The globalisation of urbanisation, as it were, with its flows of people and resources into and between cities across the world (and its corollary, the depopulation of rural areas in developing countries) is undoubtedly one of the biggest phenomena of the twenty-first century (UN DESA 2014). Add to this mix major changes in the technological based of urban development and the pressure on the environment that urban change brings provide all cities with new and continuing challenges. While all cities may be said to be facing similar challenges, the same cannot be said to be the case in terms of how they are responding to these changes, or indeed the extent to which individual cities share in the advantages or disadvantages of these economic and social challenges. However, what can be said to be the case is that cities largely fuel the global economy, both in the developed and developing world. Indeed, a key feature of the early twenty-first century is the extent to which cities in countries such as India, Indonesia, and China, for example, are shifting the economic balance of power from the developed to the developing nations. One of the consequences of this change is that not all cities are equal, either in their significance or importance in this increasingly globalised urban system. The McKinsey Global Institute, for example, has suggested that 440 of the world's 600 most economically significant cities are in developing nations, and these are projected to account for almost half of all urban growth globally by 2025 (McKinsey Global Institute 2011). This shifting power in the global urban hierarchy has meant that there is now far more competition between cities across the world for investment and resources. This has inevitably seen some cities win and others lose, some thrive while others wither, and some grow rapidly and others experience considerable economic, social, and environmental decline. The traditional narrative of this decline and rise has often been told as a story of older industrial cities declining because of stronger economic competition from cities in developing nations. This is set against the rise of developing nation cities whose growing urban economies have been based on offering cheaper modes of production or new forms of economic activity and new technological developments. What is interesting, however, is that some of those cities that have grown amidst this global urban restructuring are now also beginning to suffer. A recent report for the Royal Town Planning Institute (RTPI) in England makes this point well: 'Cities as well as nations are increasingly competing for people, investment and dominance in a particular sector. Manufacturing cities such as Liverpool and Detroit have shrunk in part due to competition from places with less expensive labour costs. These are issues not just for the developed world; for example, Majalaya, Indonesia, once known as 'dollar city' and a leader in textile manufacturing, has been hit by globalisation and increased competition from places like China using more advanced machinery' (Hubbard 2014, p.10). Such fluctuations in the fortunes of cities across the world are playing out against the backdrop of an uncertain future for global economic growth. While the recent global recession has put a hold on the long-term rates of economic growth for many developed nations, other parts of the world continue to experience growth (even if these rates of growth themselves are not as spectacular as they have been previously). However, overall the picture for global economic growth in the twenty-first century is one of slower growth than was experienced for much of the second half of the twentieth century.

This rise in urban dwelling is matched by the continuing decline in rural populations, where the near peak of 3.4 billion population (by 2020) is predicted to fall to around 3 billion by 2050. Again, the nuances of this urban to rural population change are interesting, with highly urbanising regions such as Africa and Asia accounting for approximately 90 per cent of the global rural population (UN DESA 2014).

What drives much of the growth in the significance of cities globally has been urbanisation. Currently, 54 per cent of the world's population lives in urban areas, and by 2050 this is predicted to rise by a further 10 per cent. Moreover, combining urbanisation with overall population growth is suggested as adding a further 2.5 billion people to the urban population of 2050 (UN DESA 2014), with the vast majority of this growth occurring in India, China, and parts of Africa. By 2050 India, China, and Nigeria will be responsible for 37 per cent of the projected growth of the world's urban population between 2050. The UN DESA's population division research shows that the world's urban population has grown rapidly from the middle of the twentieth century, moving from 746 million in 1950 to nearly 4 million in 2014. What it also reveals is the geography of this growth is uneven, with 53 per cent of the world's urban population in Asia, with 14 per cent in Europe and Latin America, and the Caribbean with 13 per cent. While the United Nations 2014 report highlights the continued growth of 'megacities' with their large populations and economic power, it also highlights the significance of smaller cities (those with less than 500,000 people) which are continuing to grow globally and which account for almost half of the present 4 billion urban population. The salience of this is that such cities are often those that are growing fastest and will continue to do so up to 2050 (UN DESA 2014).

With much of the growth in the urban population to 2050 predicted to be in countries in developing regions and particularly in Africa, the management of this process is critically important, along with planning for the infrastructure and facilities needed for this emerging number of urban dwellers. For John Wilmouth, Director of the UN DESA's population division, the management of this potential level of urban growth has become, '…one of the most important development challenges of the twenty-first century. Our success or failure in building sustainable cities will be a major factor in the success of the post-2015 UN development agenda' (UN DESA 2014). What lies beneath this call to greater and better management of growing urbanisation is that although such urban change often results in better opportunities and life chances for people, it can also lead to (or perpetuate) inequalities between people, cities, and regions. Data from the World Bank (Chen and Ravallion 2012) suggests that some 2.5 billion people live on less than 2 dollars a day, while nearly 3 billion people will 'enter' the middle class by 2050. This economic unevenness is compounded by the uneven impact of technological development on urban populations, where, for example, digital technologies and technological innovation has seen the growth of highly skilled and educated workers in both developed and developing countries but alongside the growth of a large unskilled service sector class of urban workers. The social and economic divisions this causes can have negative impacts on the lives of all urban dwellers with the potential for social conflict between groups split by their differential access to education and social services and their levels of income and social mobility. Such challenges to urban living are added to by the pressures that continued urbanisation places on the demand for natural resources and the natural environment, as urban growth and spread draws heavily on the land and the ecosystem services that urbanising societies eagerly consume. Greening the economy and tackling climate change are now key aspects of the management of nations and cities across the globe, and although it is fair to say that global consensus on this issue is still to be fully realised, a large number of policymakers and leaders agree that climate change is not a problem for tomorrow but is here today (Dastbaz and Strange 2015, p.3).

The energy demands then of our urbanising world are considerable, with world energy consumption predicted to grow by 56 per cent up to 2040. Although by this date renewable energy is projected to be the fastest growing energy source, carbon-based energy, such as coal, will still account for over a third of all energy supplies (Hubbard 2014, p.18). Planning for such energy use and demand is crucial if the urbanised spaces of the middle twenty-first century are to offer real benefits to their citizens (e.g. better housing, transportation, healthcare, and education) rather than potentially scarred and degraded urban environments, as a result of unregulated and unmanaged economic development. The UN DESA report on world urbanisation prospects is clear in its view that in a world of continuing urbanisation planning for cities is critical to meet the sustainable development challenges that such rapid and continuing urbanisation poses. The report (p.17/18) suggests 'that governments must implement policies to ensure that the benefits of urban growth are shared equitably and sustainably; that diversified policies to plan for and manage the spatial distribution of the population and internal migration are needed; that policies should be aimed at a more balanced distribution of urban growth; and that accurate, consistent and timely data on global trends in urbanization and city growth are required'. Ultimately, the report argues that:

> *Successful sustainable urbanization requires competent, responsive and accountable governments charged with the management of cities and urban expansion, as well appropriate use of information and communication technologies (ICTs) for more efficient service delivery. There is a need for building institutional capacities and applying integrated approaches so as to attain urban sustainability.* (UN DESA 2014, p.18)

2.3   Towards Spatial or Integrated Planning

The social and economic challenges facing cities highlighted in the previous section are being played out across global space – they are, as such, inherently spatial challenges and what some commentators call the 'critical geographies' of an emerging twenty-first-century urban complexity (Harris and Pinoncely 2014, p.6). Within this world of urbanised environments and complexity, we see the interrelated sets of issues of population change, resource depletion, environmental change, uneven economic growth, continued urban sprawl, and social inequalities playing out in old and emerging cities where despite the calls of agencies such as the United Nations, policy responses are not always clear, integrated, or informed by an understanding of how these issues impact on particular places. As Harris and Pinoncely (2014, p.6):

> For a variety of reasons, policy- and decision-making too rarely incorporate the implications of the ways in which we use land and the consequences for different places. The neglect of place, in particular the way that different policies combine to affect places in different ways, has contributed to a range of negative economic, social and environmental outcomes....

Their call for an integrated response to these challenges is a timely one and reminds of the need to respond to social and economic challenges in a way that does not see them in isolation but rather as sets of interconnected issues that necessitates a planning response. So, rather than turn away from the notion of trying to manage the changing circumstances of the twenty-first-century urban space, instead we should embrace both the theory and practice of planning, or as it is often called *spatial* or *integrated* planning:

> To this end, policy- and decision-makers can learn much from the theory and practice of 'spatial planning' (or 'integrated planning'). This goes beyond traditional land use planning to seek to integrate policies for the development and use of land with other policies and programmes which influence the nature of places and how they function. (ibid, p.7)

Such a statement takes us to two key questions that lay at the heart of debates on managing and planning the twenty-first-century cities – namely, what do we mean by planning and what should planning be about? Normatively, planning is about *'anticipating the future and attempting to shape it for the good of society'* (Ellis and Henderson 2014, p.6). As Ellis and Henderson point out, planning is carried out everywhere. So, although its early nineteenth-century origins inflect planning with views about managing the industrial city and coping with the demands of Victorian industrial modernity, planning, and the need to plan for the future, is something that is experienced (and needed) globally. Traditionally, planning has been thought of as simply about the allocation and use of land and the degree to which it can be used by different interests in different ways. Often this form of *zonal planning* of land uses has come to be enshrined in statutory legislation that produces a state-managed system of development control and development planning, supported by a formal regulatory framework of procedures, protocols, guidance, and laws. More recently, this conceptualisation of planning has been widened to include the notion of planning that seeks to take more account of the social and economic dimensions with which it deals and as such is seen to produce planning that recognises more the interconnected nature of the issues that it is faced with addressing.

What is interesting is to examine how some of these ideas around the relationship between socio-economic change and planning have been (and are being) played out in different places. In Europe, the development of the European Spatial Development Plan (ESDP) approach has offered some important pointers for planners, with its articulation of how EU-wide territorial development should be refocused to include a wide range of social, economic, and cultural issues. Many European countries have made a similar case for their national level planning agendas, such that national spatial objectives are often aligned with debates about national culture and identity and national cultural assets.

In the UK, planning has undergone a re-evaluation of what it means to do planning, shifting from being about land use towards being characterised by spatial planning (Davoudi and Strange 2009). Spatial planning is hard to define; there was no single definition of what it meant to do spatial planning. Rather there were a broad set of themes and ideas related to planning as being a collaborative, participative, integrative, and visionary process. This has led to a vague understanding of what spatial planning was within the planning profession (Davoudi and Strange 2009; Tewdwr-Jones et al. 2015). However, the arrival of 'spatial planning' in the UK in the mid-2000s, with its focus on planning beyond traditional land use, certainly offered the opportunity for 'culture' to become part of what planning did and what planners planned for. However, within a UK context, the formal acknowledgement of spatial planning as *the* approach to planning was short-lived. Following the arrival of the Conservative-Liberal Democrat coalition government in 2010, the planning system came under scrutiny as part of the government's localism agenda, a process characterised by attempts to remove and/or reduce central government control. Here, the planning system was conceptualised as being cumbersome, bureaucratic inefficient, and slow, as well as being insensitive to the needs and development aspirations of local neighbourhoods, communities, and developers. The result was the introduction of the National Planning Policy Framework (NPPF) in 2012. The NPPF was designed to rectify these criticisms, with an intention to 'simplify' the planning system making it less complex and more accessible to the public and developers. The NPPF brought with it a number of changes to the way the planning system was to operate and function in England (not least the arrival of neighbourhood planning), reducing thousands of pages of extant guidance and planning documentation into 52 pages, and subsequently added to with additional planning guidance in 2014. These changes have also been part of a broader process of 'localism' and a general movement towards the decentralisation of power(s) from central government towards local government and local communities and neighbourhoods (Brownhill and Bradley 2017).[1]

2.4   Dimensions of Planning for the Twenty-First-Century City

In this section, I want to return to some of the key contemporary spatial challenges that face cities in the twenty-first century and how planning and planners are (or might be) responding to those challenges. While it is recognised that there are many

---

[1] This paragraph draws on material previously written for Strange (2016b).

issues that one could consider, I want to focus on the following set of challenges – those that are loosely based on the UN DESA world urbanisation prospects discourse outlined earlier in the chapter. These are the issues of economic unevenness, digital infrastructure and technology, environment and resources, social justice and equality, and urban governance.[2] I will now consider briefly each of these issues and the actual and potential response of planning and planners to the threats they pose to sustainable urban living.

2.4.1 Uneven Economic Development

In his paper *Creating Economically Successful Places*, Jim Hubbard argues that:

> The global economy is changing faster than ever. Increased competition, the rise to prominence of economies such as China and India, technological change, flows of financial capital and other factors mean that existing industries and old certainties are being challenged as never before .... All this means that, in the twenty-first century, what makes an 'economically successful place' goes well beyond a narrow approach to development. Planners have a critical role in helping communities at all levels prepare for and navigate these challenges. (Hubbard 2014, p.6)

Planners do indeed have a role to play in shaping the future of rapidly changing urban circumstances. But what might be some of their options to tackle the often uneven ways in which such change is taking place, and how can cities plan effectively in the wake of shifting global capital and policy environments that can be governed by national rather than local political agendas? Hubbard's RTPI paper is instructive here in that it offers several clear suggestions and examples of where the role of planning and planners working in the face of such changes could be effective. For example, quality of life indicators plays a major role in determining the success of cities. Indeed, technological change has resulted in both the need for fewer workers and work hours in cities and also different types of works – one consequence of this has been a heavier demand for spaces in cities which are designed to accommodate activity related to 'leisure time' that is away from work. Those cities that can demonstrate built environments that cater for such activity and which have good quality open space are seen to offer better opportunities and improved health and well-being for its citizens. For example, cities such as Melbourne, Vancouver, and Toronto are noted for the provision of quality public spaces and are typically at the top of urban liveability indexes (ibid, p.22). Hubbard also outlines the ways in which a longer-term economic view of cities is also significant for economically successful futures (ibid, p.30). He draws firstly on the example of Dublin's expansion in the late twentieth century, a growth process fuelled by a credit-led construction boom and ending in sclerotic recession following the credit

---

[2] There are, of course, many issues that could be covered in this discussion. I have chosen those that seem to me to be the most critical issues and those to which planners and planning may have something to contribute in the way of a coherent response.

crunch in 2008 and the collapse of Dublin's urban property market, and secondly on contemporary changes in retailing which means that both out-of-town and high-street retailers are feeling the effect of new online shopping habits (let alone the previous effect of out-of-centre shopping malls on the older traditional high streets of the UK towns and cities in the 1980s and 1990s). While insightful, much of this analysis and commentary is based on an acceptance of a prevailing neoliberal discourse that has come to surround interpretation of urban change and the globalisation of urban policy in the twentieth and early twenty-first century (Cochrane 2007). As Lees et al. (2016) make clear, while the neoliberal discourse of economic restructuring and its impact on cities is pervasive, such impacts are felt differently in different places. Hence what matters is how cities respond to the wider process of economic change, what choices they make about how to respond, and to avoid reading off such responses as the inevitable consequences of neoliberal economic urban policy (p.116).

2.4.2   Digital Infrastructure and Technology

> *Our urban future demands innovation. cities are at the front-line in responding to global challenges of resource scarcity, climate change, unemployment and ageing populations. while these are big challenges, they also present major new business and innovation opportunities… Demand is significant: cities across the world continue to grow and the global market for integrated urban solutions is estimated to reach £200 billion by 2030. More widely, it is estimated that at least Us$40 trillion will need to be invested in urban infrastructure worldwide over the next 20 years.* (Catapult Future Cities 2016, p.1)

The Future Cities Catapult is a UK government-funded programme that adopts a city systems approach to solving complex urban problems; the programme is predicated on the idea that an integrated approach to the solution of these urban problems is an essential starting point for successful city futures:

> *Urban problems are the result of multiple factors with far-reaching impacts involving complex feedback loops. Given the complexity of urban problems, the most effective approach to resolving them considers a city's multiple systems simultaneously, rather than focusing on how to fix a particular element. A city systems approach is just this; it considers the city as a system and designs solutions to have maximum positive impact while minimising negative unintended consequences.* (Catapult Future Cities 2016, p.1)

The multidisciplinary approach to urban planning that this entails is critical in bringing together urban planners, engineers, and data and social scientists to work on the master planning of cities. Although it is UK based, the *Catapult, Future Cities* programme is a good example of how new technology and digital software and hardware are being brought together in the remaking and envisioning of urban space. Drawing on expertise in the digital creative industries and data visualisation and modelling, the objective is to demonstrate for urban planners and policymakers (wherever they may be) that access to such skills and expertise can produce better cities now and in the future. Outcomes, although infused with technology, are said to have people or a 'human-centred design' at its core (ibid, p.3). An innovative example here includes the development of *City Mapper*, a smartphone app that offers journey planning using real-time public transport data. Developed first in Greater London, the app is now used by millions of individual daily travellers in cities such as Paris, New York, and Berlin. The catapult programme claims that the app is improving transport supply and demand in these cities and increasing transport capacity significantly at less expense than building often intrusive and disruptive transport infrastructure. A similarly innovative project is the development of *MKSmart*, a citywide 'Internet of things' project in Milton Keynes that is experimenting with new technology for connecting devices such as '...car parking, pest control and waste disposal' (ibid, p.7).

While many benefits can be brought about because of technological innovation, the rapid pace of technology-based development can also bring changes to, for example, patterns of urban work and employment (Sabri et al. 2015). This is the case for both developing and developed regions through increases in productivity brought about by the adoption of new technologies (see, e.g. Hubbard 2014, p.14). There will be both job losses and gains in this process – to tackle each of these impacts of new technology require cities to focus on the education and training of their populations, reskill its urban populace such that they can show resilience in the face of technological transformation of the urban environment and work spaces that such change can both create and destroy. Indeed, technological change and its impact on urban economic activity can be significant for city success. Hubbard (2014) shows how for European cities a city's resilience and ability to adapt to technological change (and hence its long-term economic competitiveness) is related closely to its ability to diversify its economic base, its rate of technological innovation, its accessibility, and its overall connectivity (p.14).

The reshaping and reimagining of the urban environment has been opened up by the introduction of new digital forms of technology that have not had such potential impact on the shape of cities since the beginning of industrialisation. The traditional planning of cities based on industrial patterns of employment and land use will become (and are becoming) outmoded. New spaces of the city are being developed in response to both the decline of industrial forms of production and the rise of digital infrastructure, changing, for example, the complex relationship between land, buildings, and the people who use them. As Hubbard states 'The use of buildings and land will become more complicated as further innovations become commonplace, but these changes may also enable more sustainable and liveable communities where people are able to travel shorter distances to meet everyday needs' (ibid, p.15).

2.4.3   Environment and Resources

One of the major challenges facing urban planners as the world continue to urbanise is the extent to which cities can cope with the consequences of climate change and environment and resource depletion. We know that global temperatures are rising and that many regions are becoming hotter and drier. Cities in California and Australia have, for example, recently experienced their hottest years on record, while the Summer of 2017 saw a heat wave sweep across parts of Southern and Eastern Europe to an extent not seen before. We know that along with such climate events come large numbers of premature deaths from those suffering the effects of such severe heat (Wootton and Harris 2014), and that such events are felt most acutely in cities. In urban areas, the effect of excessive heating (the urban heat island phenomenon) has major implications for the heath or urban residents and considerable impact on the economic cost of work days lost. The role of planning here is complex, for example, in changing travel to work patterns and promoting alternative sources of energy that can help in the ways that cities might mitigate and adapt to changing climatic conditions. In Tokyo, for example, its master-planning programme is working to reduce the city's urban heat island by designing public infrastructure that reduces urban wind flows and absorbs heat and moisture (Wootton and Harris 2014, p.11).

Other climate effect events are occurring. Sea levels are rising with major implications for planning including the loss of coastal towns and coastal communities. Hurricanes and major storms are becoming regular occurrences and ones that often devastate urban infrastructure, affecting those from wealthy and poorer regions indiscriminately. Climate research suggests that levels of rain fall are increasing and that flooding is a major threat to cities around the world. Wootton and Harris (2014) provide data which suggests that in England the current one million properties at risk of surface flooding will quadruple by 2080, with resulting damage to properties and buildings of around £12 billion per year. They also add that such flooding poses a major risk to large amount of urban transport and energy infrastructure (ibid, p.12). Significantly, they highlight evidence that suggests a strong link between flooding and weaker urban economic growth, citing a study on European countries that found a correlation between increases in an areas rainfall and a 1.8 per cent reduction in that areas GDP growth (ibid, p.12). Mitigating the effects of flooding is thus a major challenge for many urban planners. However, many cities can point to examples of novel building techniques, the introduction of drainage plans and the restoration of upland landscapes that have had a positive effect on reducing flooding and its localised impacts. Others have developed strategies for improving their urban ecosystems with planning becoming a key element in the creation of more sustainable forms of built environment (Mersal 2016). Similarly, the 'greening of cities' though lower-energy consumption is becoming something of a mantra for planning in cities of the developed nations (Andersson 2016) as well as those developing nations where cities are continuing to urbanise, such as in India and China. The use of planning policy then is key, and its role is essential in helping us to move towards a carbon-reduced society. As Wootton and Harris (p.17) argue, planning has a role to play by:

> ...informing the density of development and its location, how it is integrated into other land uses and the location of renewable energies, and helping to promote walking, cycling and public transport and dis-incentivise private car use. Planning systems are one means to specify the design and energy efficiency of new and existing homes and buildings....

2.4.4 Social Justice and Equality

One of the enduring features of urbanisation is the persistence of poverty and spatial inequalities in cities throughout the world. Alongside the many examples of economic growth and success of cities, there are as many examples of urban poverty and urban decline. We find such cases both in cities in the developing and developed world where shifting patterns of economic production, along with often poorly managed responses to change, have resulted in social inequalities experienced by millions of people around the world. Whether these inequalities are experienced as poor-quality housing, lack of access to amenities and resources, limited access to (often low skilled) employment opportunities, or living in densely populated areas in overcrowded conditions, the reality for many urban dwellers is one of poverty, a low standard of living and limited opportunities for personal or familial betterment. Cities have of course always been places of extremes of wealth, affluence, and opportunity. For some this is the consequence of the process of capitalist urbanisation, while for others it is simply an outcome of the natural feature of urban environments. Whatever the perspective, and even with a cursory knowledge of the history of planning, much of planning's *raison d'etre* is about how to respond to the consequences of this uneven development and the spatial effects of the twin processes of industrialisation and urbanisation. The early history of planning has its roots in tackling social problems brought about by industrial and urban growth in the nineteenth-century city and its attempts to make those cities better places in which to live. However, the continuing presence of social ills and problems in cities suggests that despite the considerable successes of planning in dealing with some of these social problems through the implementation of policies design to remedy the spatial manifestation of uneven development, successive waves of urban policy and planning across the world have not prevented the continuation of deeply rooted social and economic problems.

What is clear in this debate on social justice and equality is that *place matters* to peoples' life chances. While it is perhaps not universally agreed, it is probably safe to suggest that *where* one is born or lives is intricately linked to *how* an individual is able to live a fulfilling and productive life (Dorling 2001). Or, to put it another way, 'Place, as much if not more than individual characteristics, can undermine people's ability to live the best lives they possibly can' (Pinoncely 2014, p.8). While not the only factor to consider, access to housing is one of the major indicators of inequality in cities, and as such is a key driver of the continuation of place poverty. Access to housing has thus been a major site of social unrest in cities, whether that be protest against the neighbourhood cleansing and gentrification of city centres or the social unrest of young people on the (often) peripheral housing estates of larger urban centres. Indeed, despite impacting differentially across the globe, gentrification is a key maker of places through its remodelling of spaces of the city in favour of an urban middle class. As Lees et al. argue in their book on *Planetary Gentrification*, 'We show that gentrification is a phenomenon that cities worldwide have experienced (it is not totally new in the twenty-first century to the global South) and are experiencing (through different types of urban restructuring)' (Lees et al. 2016, p.5). More broadly, there is no shortage of examples of social protest and resistance to urban renewal and restructuring. Cities in Europe, the United States of America, and China, for example, have all experienced urban protests as local populations challenge prevailing governance regimes over issues such as pollution, access to resources and amenities, or the privatisation of public space.

So, what role does planning play in this process? Of course there is planning's traditional role as a mediator over the use of resources and as a key player in the allocation of land and the spatial distribution of goods and resources. This is a set of functions that one might hope (normatively) be continued. However, the reality is that such approaches are not universal and even where they are in evidence spatial policies are not always successful. However, it may also be that planners and planning (in its widest sense) have new opportunities for tackling some of the spatial outcomes of economic and social change. Pinoncely (2014, pp.3, 4), in her discussion on poverty and place, identifies some key ways in which planning can provide better ways to tackle poverty and inequality. Perhaps the most significant of these for promoting social justice in cities is that planners work at:

- Integrating poverty reduction strategies tailored to their particular places and communities that need to be developed;
- Recognising the importance of planning in poverty reduction within their local plans;
- Using more localised approaches to planning (such as neighbourhood planning in the UK) as tools for helping to improve communities and reduce poverty local;
- Developing and implementing better data relating to poverty and inequality that reflects the importance of place and the environment.

2.4.5   Urban Governance and Planning

The challenges that urban areas face whether in the developing or developed regions require a response that is effectively a managed intervention into the built environment. This means that policymaking and decision-making requires effective structures that allow for coordinated and integrated responses. In other words, planning in and for the twenty-first-century city requires governance. At a global scale, the governance of cities is enacted in a very different ways, from top-down hierarchical modes of city governance to more localised or democratic and federated systems of urban decision-making. As with most of the ways in which planners might respond to the challenges of continuing urbanisation, there is clearly not going to be *one* form of governance structure or institutional apparatus that works. However, if cities are to effectively manage and deliver outcomes which result in better places for their citizens, then more democratic forms of management and intervention offer the most likely ways of addressing the core development issues that many cities face. What this will entail is more collaborative forms of planning (Healey 1997), where planners work with other stakeholders bringing together cooperative efforts to produce strategically informed plans whose policies are integrated responses to key urban concerns such as housing supply and affordability, economic development, and transport and urban connectivity. Some countries have developed approaches to planning where strong institutional and stakeholder relationships work to produce clear integrated and strategical spatial plans (cities in Germany and the Netherlands offer good examples here) (Harris and Pinocely 2014, p.36; see also Davoudi and Strange 2009). Other countries and cities may require more institutional development to allow such forms of collective and strategic planning that includes 'establishing stronger legal and regulatory frameworks within which organisations, institutions and agencies operate, better land use policies and support for local organisations and residents to help shape the development of their communities' (Harris and Pinocely 2014, p.39). Such changes may be more appropriate in developing countries where there is often a stronger need for institutional support for planning, but this is not exclusively the case.

For example, the English planning system, where planning has largely been centrally determined but locally implemented, is currently experimenting with its own form of greater local accountability in the form of localism and neighbourhood planning (Davoudi and Madanipour 2015; Brownhill and Bradley 2017). Changes brought about as part of the arrival of the National Planning Policy Framework (see previous discussion in this chapter) have also been part of a broader process of 'localism' and a general movement towards the decentralisation of power(s) from central government towards local government and local communities and neighbourhoods, enshrined in the Localism Act (2011). Whether one sees such changes as empowering local communities in the local development process, or as a form of neoliberal social regulation (see Brownhill and Bradley 2017, p.21 for a debate) offering the illusion of local power in the wake of the market, the localism agenda in the UK is an important departure point in contemporary discussion around the role of local communities in shaping their development futures. The turn to localism has most recently found expression in the form of an urban devolution agenda being pursued by the UK Conservative government and is partly about an attempt to 'rebalance' England's economy and strengthening the English north and its deindustrialised cities. The full effect of devolution deals on English cities is yet to be seen. It may well be the case that the whole process runs out of steam as national political agendas change, but city-focused devolution raises important issues for the planning, not least in terms of the role cast for planning as a key agent in local and national growth and recovery agendas.

So, what does current debate suggest might happen in a devolved environment to planning?[3] I have previously written that there are four implications for planners here.[4] Firstly, planning cannot be a completely devolved activity – supralocal forms of networked governance will be necessary in any new devolved (financial or political)

---

[3] For a more extensive debate about the role of planning in developing city futures in a devolved context, as well as the role of universities in this process see Tewdwr-Jones et al. (2015).

[4] The issues raised in this paragraph have been previously considered in Strange (2016a).

system. Secondly, planning in a devolved context will need to be more spatial, with better development and use of local intelligence data and information. Thirdly, effective planning outcomes will probably be those that have been actively promoted both city-region wide *and* by local community engagement processes. Planners will thus have to be able to work at multiple scales in ways that, while part of their present role, will be significantly enhanced. And fourthly, planning and planners will need to better understand and analyse the interrelationships between different sectors of the economy, particularly focusing on how socio-economic processes that occur locally might be connected to broader extra-local patterns of development and change (Strange 2016a, p.19).

2.5   Conclusion

In the introduction to this chapter, I suggested that there were key questions that the chapter would seek to address. These were: How do cities, and those who plan for them, respond to these challenges so that urban futures produce fairer and more just places? What are the technological challenges that face urban planners and decision-makers as they search for solutions to complex and multi-faceted urban problems while cities transition from their analogue pasts to their digital futures? What forms of governance and local democracy allow greater degrees of citizen involvement and participation in the making of urban futures? And, what is the role of the state in shaping the policy and political context within which cities can plan for their futures?

The preceding discussion has sketched out some of the responses to these questions and issues, but they are difficult ones to answer and will continue to occupy those charged with providing a response to the march of urbanisation for many years to come. The economic and social challenges that face cities over the next 50 years are, as has been outlined, considerable. To be able to respond effectively to such challenges, planned responses need to be ones that are informed by evidence and data on the key trends of urban development such that decision-making can be informed and related to the actual experiences of places. Equally, they should be situated within an integrated approach to the planning of place and space, where the skills and expertise of a wide range of interests and stakeholders are brought to bear in tackling those difficult and complex issues. Responses need to be informed by an understanding of the nature of the characteristics of the cities and their diverse communities and how those places are responding to changing circumstances in their economic and social bases. The responses need not only to be technically informed but also democratically determined, maintained, and communicated to all within the city. Planners are (potentially) well placed to fulfil these roles and communicate to wide audiences the difficult choices they must make as they work on the distribution of scarce resources.

Urban policymaking needs to be informed by the complexities of space and place and the operation of economic, social, and environmental forces in particular places. Policymakers need to be more spatially informed, or what some commentators have called the need for greater *spatial thinking and action*, where policymaking is informed and contextualised within a better understanding of place. As Harris and Pinoncely (2014) argue:

> A much greater spatial awareness and intelligence will improve the decisions that are made, and the consequences for the everyday lives of people and communities. Further, spatial thinking might help to counter a commonly expressed concern about contemporary politics and policy found in many countries – the sense that too many of our political leaders and decision-makers lack long-term visions for change to produce a better society for all. (p.35)

Thinking and acting spatially will be no easy task. It has been a difficult switch for planners as they try to turn away from a conception of planning that is just about the location and use of land and move towards a more holistic and integrated approach. It is perhaps even more challenging for local and national politicians as they work to establish their visions for the future of their towns and cities. Planners have a major educative role here in their offer of knowledge and information to policymakers about the relationships between space, place, and future economic and social well-being in ways that can pave the way for competitive advantage (Hubbard, 2014, p.40).

In their book, *Rebuilding Britain: Planning for a Better Future,* Hugh Ellis and Kate Henderson make a convincing case for the importance of planning to society and the degree to which 'morally and practically' town planning is essential in helping meet the major economic, social, and environmental changes of the early twenty-first century (Ellis and Henderson, 2014). While they chart the development of town planning from the years after 1945, their claim for the contemporary salience of planning is a call to connect with some of the views and visions of the nineteenth-century planning movement and particularly making places better for people as inscribed in the idea of, and drive for utopia, that inspired much of modernist planning thought. For these authors, it is this 'idea' of utopia that has been lost in contemporary planning, not just in the Great Britain, but globally. Reconnecting with the utopian traditions of planning's seminal thinkers and practitioners is key if one of the fundamental objectives of planning it to be reignited – its pursuit of building better places. In reconnecting with this faded planning idea, we will be reconnecting with an idea that is very much needed for the planning of our cities into the future (Ellis and Henderson 2014, p.139). Ellis and Henderson's view may be one that is difficult to reach, but as we continue to see cities facing complex and spatially entrenched challenges, their call is one to which we would be wise to listen. As they say 'The real task is to draw all these solutions together and see what that might offer by way of a new society. This was the vision that those, such as Ebenezer Howard and William Morris, set out over a century ago and it is an extremely hard task' (Ellis and Henderson 2014, p.139).

References

Allmendinger, P. (2011). *New labour and planning. From new right to new left*. London: Routledge.
Andersson, I. (2016). 'Green cities' going greener? Local environmental policy-making and place branding in the 'Greenest City in Europe'. *European Planning Studies, 24*(6), 1197–1215.
Brownhill, S., & Bradley, Q. (Eds.). (2017). *Localism and neighbourhood planning. Power to the People?* Bristol: Policy Press.
Catapult Future Cities. (2016). How can the UK innovate for the world's cities? Catapult Future Cities, UK Trade and Investment, London.
Chen, S., & Ravallion, M. (2012, March). *'An update to the World Bank's estimates of consumption poverty in the developing world*. New York: The World Bank.
Cochrane, A. (2007). *Understanding urban policy*. Oxford: Blackwell.
Dastbaz, M., & Strange, I. (2015). Sustainability, design and the built environment: Context, current agendas and the challenges ahead. In M. Dastbaz, I. Strange, & S. Selkowitz (Eds.), *Building sustainable futures: Design and the built environment*. Basel: Springer.
Davoudi, S., & Strange, I. (2009). *Conceptions of space and place in strategic spatial planning*. London: Routledge.
Davoudi, S., & Madanipor, A. (2015). *Reconsidering localism*. Abingdon: Routledge.
Dorling, D. (2001). Anecdote is the singular of data. *Environment and Planning A, 33*, 1335–1369. 18.
Ellis, H., & Henderson, K. (2014). *Rebuilding Britain. Planning for a better future*. Bristol: Policy Press.
Ellis, H., & Henderson, K. (2016). *English planning in crisis*. Bristol: Policy Press.
Harris, M., & Pinoncely, V. (2014). Thinking Spatially, RTPI Planning Horizons Paper No.1, London RTPI.
Haughton, G., Allmendinger, A., Counsell, D., & Vigar, G. (2010). *The new spatial planning: Territorial management with soft spaces and fuzzy boundaries*. London: Routledge.
Healey, P. (1997). *Collaborative planning*. London: Palgrave Macmillan.
Hubbard, J. (2014). *Creating economically successful places*. RTPI Planning Horizons Paper No.4, London RTPI.
Lees, L., Shin, H. B., & Lopez-Morales, E. (2016). *Planetary gentrification*. Cambridge: Polity Press.
McKinsey Global Institute. (2011, March). *Urban world: Mapping the economic power of cities*. New York: MGI.
Mersal, A. (2016). Sustainable urban futures: Environmental planning for sustainable urban development. *Procedia Environmental Sciences, 34*, 49–61.
Pinoncely, V. (2014). *Promoting Healthy Cities*. RTPI Planning Horizons Paper No.3, London RTPI.
Rydin, Y. (2013). *The future of planning*. Bristol: Policy Press.
Sabri, S., Rajabilfard, A., Ho, S., Namzi-Rad, M., & Pettit, C. (2015). *Alternative planning and land Administration for Future Smart Cities*. IEEE Technology and Society Magazine, December, pp.33–35.
Strange, I. (2016a). Planning and Leeds City Region: Some issues for planning as a devolved public sector service. In J. Shutt (Ed.), *"Until we have built Jerusalem". The role of universities in the changing Northern Political and Spatial Geography* (pp. 18–19). Leeds: Leeds Beckett University.
Strange, I. (2016b). Urban planning, architecture and the making of creative spaces. In P. Long & N. Morpeth (Eds.), *Tourism and the creative industries: Theories, policies and practices*. London: Routledge.
Tewdwr-Jones, M., Goddard, J., & Cowie, P. (2015). *Newcastle city futures 2065: Anchoring universities in cities through urban foresight*. Newcastle Institute for Social Renewal, Newcastle University, Newcastle.
Wills, J. (2016). *Locating localism. Statecraft, citizenship and democracy*. Bristol: Policy Press.
Wootton, G., & Harris, M. (2014). *Future-Proofing Society*, RTPI Planning Horizons Paper No.2, London RTPI.
UN DESA. (2014). *World urbanization prospects*. New York: UN DESA's Population Division notes.

Chapter 3
Urbanisation and Entrepreneurship in Development: Like a Horse and Carriage?

Wim Naudé

3.1 Introduction

In the process of development, urbanisation and entrepreneurship go together, like a horse and carriage. But like love and marriage, the relationship is not a simple one. Both urbanisation and entrepreneurship are central 'demographic facts' of the early twenty-first century. No economy has developed without urbanising (Zhang 2002). The majority of the world's population already resides in urban areas. And the rest is continuing to join them at a fast rate. As Ovanessoff and Purdy (2011) note, the urban population in the developing world is set to 'more than double between now and 2050 to 5.3 billion' (p. 46). The number of megacities continues to grow. Each year, around ten cities the size of New York comes into being in the developing world. Likewise the number of entrepreneurs, defined as those who are self-employed, continues to increase: there are at least a billion entrepreneurs now in the world (Naudé et al. 2017). Most entrepreneurs are to be found in these cities and megacities.[1]

Entrepreneurs hasten urbanisation, but urbanisation also benefits entrepreneurs, so that a virtuous cycle comes into being.[2] As Fig. 3.1 shows, there is a strong positive correlation between how entrepreneurial a country is (measured by its score on the Global Entrepreneurship Index) and its level of urbanisation (measure by the share of the population that is urbanised).

---

[1] For instance, a study from Sweden found that start up rates of new firms per 10,000 population where 32% higher in cities (metropolitan municipalities) than the national average (Olsson et al. 2015).

[2] In the case of the USA, Glaeser (2007) concludes that 'more entrepreneurial cities are more successful'.

W. Naudé (✉)
Maastricht School of Management and Maastricht University, Maastricht, The Netherlands

IZA Institute of Labor Economics, Bonn, Germany
e-mail: Naude@msm.nl

© Springer International Publishing AG, part of Springer Nature 2018
M. Dastbaz et al. (eds.), *Smart Futures, Challenges of Urbanisation, and Social Sustainability*, https://doi.org/10.1007/978-3-319-74549-7_3

29

Fig. 3.1 Urbanisation is associated with better entrepreneurship (Source: author's compilation based on data from the World Development Indicators, World Bank and the Global Entrepreneurship Development Index 2017)

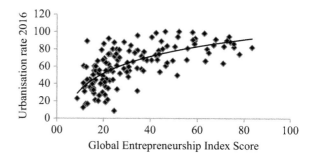

*(Source: author's compilation based on data from the World
Development Indicators, World Bank and the Global
Entrepreneurship Development Index 2017)*

Fig. 3.2 Urbanisation is associated with higher GDP per capita (Source: author's compilation based on data from the World Development Indicators, World Bank)

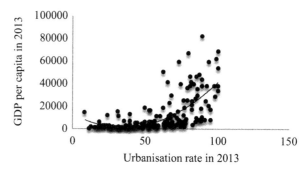

*(Source: author's compilation based on data from the
World Development Indicators, World Bank)*

The ensuing process of structural transformation is one of the salient facts of development, wherein entrepreneurs create jobs in urban areas, in new sectors such as industry or services, attracting labour and other entrepreneurs out of the agricultural sector and out of rural areas. Eventually, if these developing urban areas and their entrepreneurial dynamics can be appropriately managed and incentivised, it may lead to improvements in GDP per capita (Fig. 3.2) and many other improvements in the quality of life, such as, for example, reduced deaths from communicable diseases, maternal, prenatal and nutrition conditions (Fig. 3.3). Glaeser (2011) has called cities the world's 'greatest invention'.

In this chapter, I discuss the relationship between entrepreneurship, urbanisation and development over the various phases of urbanisation that a stylised country will go through. First, in Sect. 3.2, I set out a simple model of entrepreneurship and urbanisation; to illustrate that without entrepreneurial innovation and the movement of labour from the countryside to the city, a society can get stuck in a situation of rural stagnation.

Then in Sect. 3.3, I discuss some of the potential pathologies that can obstruct sustainable developed in the early phases of the entrepreneurship-urbanisation

Fig. 3.3 Urbanisation is associated with reduced deaths from communicable diseases (Source: author's compilation based on data from the World Development Indicators, World Bank)

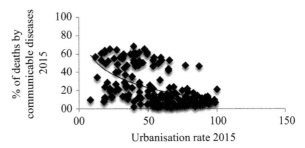

(Source: author's compilation based on data from the World Development Indicators, World Bank)

relationship. In Sect. 3.4, I continue by describing how successful cities become 'entrepreneurial hotspots', 'global start-up cities' and even 'smart cities', leading to an explosion of services consumption goods but also tensions due to competition, price rises and negative environmental impacts.

Finally, as I set out in Sect. 3.5, there comes a stage when more and more people residing in cities, the imperative to address environmental concerns and vulnerability to shocks become more important, and if well managed and regulated, more and more entrepreneurs enter the circular and sharing economy, contributing towards sustainable development.

Section 3.6 concludes, pointing out that given that urbanisation is a relatively recent phenomenon in human history. In 1800, only 3% of the world's population was urbanised (Bezerra et al. 2015). The most interesting period in the world's urbanisation is thus still to come. Entrepreneurs will for sure play a central role in this future.

3.2   A Model of Entrepreneurship, Rural Stagnation and Urbanisation[3]

Gries and Naudé (2010) proposed a stylised model of entrepreneurship and urbanisation. In this section, I summarise and present the essential elements of their model, which shows how entrepreneurship (and innovation) or its lack, can lead to the process of urbanisation and development being set in motion, or not. I discuss some of the comparative statics generated by the model, such as that entrepreneurial ability and business conditions, including financing, are essential factors or conditions that drive the rise of cities.

---

[3] This section is taken from Naudé (2016a).

3.2.1   Rural and Urban Sectors

Gries and Naudé (2010) start by noting that in underdevelopment, countries are characterised by a large and growing rural population. They denote the rural population in a representative underdeveloped country by L number of people earning an income w and Δ_T earning no income. These latter people provide what has been termed surplus labour (Lewis 1954). Total population in the rural economy is

$$\text{Pop}_T = L + \Delta_T = L\left(1 + \delta_T\right), \quad \text{with} \ \ \delta_T = \Delta_T / L \tag{3.1}$$

where δ_T denotes the ratio of *surplus labour* to income earners in the rural economy.

Population growth in rural areas is assumed to be a function of per capita income in the sector, y_T, and can be expressed as:

$$\gamma_L \equiv \frac{\dot{L}}{L} = g_L\left(y_T\right) = y_T^\varphi \ \ \text{with} \ \ \varphi < 0 \tag{3.2}$$

φ denotes the elasticity of population growth with respect to per capita income in the rural areas. This indicates a negative relationship between income per capita in rural areas and population growth.

More specifically, it can be shown that population growth in the rural sector is a negative function of the rural unemployment rate (labour surplus rate) δ_T (and a positive function of the marginal and average labour productivity (a_T)):

$$\gamma_L = \left(\frac{a_T}{\left(1 + \delta_T\right)}\right)^\phi$$
$$= g_L\left(\delta_T, a_T, \phi\right), \ \ \text{with} \ \ \frac{dg_L}{d\delta_T} < 0, \frac{dg_L}{da_T} > 0, \frac{dg_L}{d\phi} > 0 \tag{3.3}$$

At any given moment, the unemployed (surplus labour[4]) will be searching for either wage employment or self-employment on farms (as farmers) or be searching for opportunities to start-up an entrepreneurial venture (as a small firm) in the urban sector, i.e. in secondary cities and/or primary (capital) cities. If the start-up is successful, the respective agent will leave the rural areas and migrate to the urban area. Hence, migration out of farms and out of farming areas is critical for urbanisation.

The urban sector consist of large formal firms and small, informal firms. Large formal firms are typically owned either by foreigners (multinational firms) or the government. These firms produce final goods (Y) for consumption in both urban and rural areas using human capital H and N intermediate inputs x_j, which they will

---

[4]As in the Lewis model, the assumption of surplus labour means (disguised) unemployment in rural areas and on farms.

outsource to the N small household enterprises. The production function for the representative large formal firm producing for the final goods market is

$$Y = AH^{1-\alpha} \sum_{j=1}^{N} \left(x_j\right)^{\alpha} = AH^{1-\alpha} Nx^{\alpha} \tag{3.4}$$

In Eq. (3.4), A is a technology shift parameter. Producers of the final good maximise profits according to the profit function $\pi_Y = Y_M - w_H H - Np_j x_j$ with p_j the price of intermediate good x_j and w_H denoting the returns on the entrepreneurial and managerial abilities of the owner of the formal firm. Using the first-order conditions, the demand for each intermediate (service) input, is

$$x_j = H \left(\frac{A\alpha}{p_j} \right)^{\frac{1}{1-\alpha}} \tag{3.5}$$

The demands for these j intermediate inputs constitute an opportunity for the people in rural areas, Pop_T. Not all intermediate goods demanded constitute an opportunity, as some will already be taken by other household enterprises. The number of opportunities, O, in the urban area is the number of opportunities for subcontracting to a formal large firm that has not yet been taken by existing household enterprises and can be written as

$$O = \delta_N N \tag{3.6}$$

where $\delta_N = \Delta_N/N$ with Δ_N the number of offered but yet not realised business opportunities.

If they spot these opportunities, O, households will form an enterprise in an urban area. Each will then produce a unique intermediate input that they sell to the large formal firms (in this Gries and Naudé (2010) follow Ciccone and Matsuyama 1996). The requirement that they each produce a unique input x_j allows Gries and Naudé (2010) to introduce innovation, start-ups and start-up costs in the urban economy into the model and hence ultimately deal with the obstacles in the process of urbanisation.

3.2.2   Entrepreneurs, Innovation and Household Enterprise Start-Ups

If a household in rural areas spots an opportunity to provide a unique intermediate input x_j to established large firms in the urban areas, it needs to create a new household enterprise. This involves incurring costs to differentiate the product to the needs of the large firm and to finance the start-up and running costs of the household enterprise.

Gries and Naudé (2010) assume that start-up costs will depend on the density of already existing household enterprises. The more competition there is, the more costly it is to break into the market, including migrating to a more congested non-rural destination. If there is N-number of household enterprises operating in the modern economy, and total output (GDP) in the economy is denoted with Y, then I can write the density of household enterprises as N/Y. Start-up costs as a function then is $\chi(N/Y) = \varepsilon\dfrac{N}{Y}$, with $\varepsilon > 0$. In addition to start-up costs, there are permanent operational costs to operate the household enterprise. These costs are denoted by c_x.

These start-up costs need to be financed from an external finance because typically as the rural households do not have sufficient own assets from which to finance this.

The decision of the rural household on starting a household enterprises or not depends on the expected profits in comparison to agricultural wages. Once the household enterprise has been started, profits are the difference between its sales (price times quantity) and operating costs, which is

$$\pi_j^x = \left(p_j - c_{xj}\right)x_j \tag{3.7}$$

Since the prospective household enterprise is assumed to have no immediate income or accumulated savings, the start-up costs must be financed and at a loan rate of r_1. Given that the financial sector is often not very well developed in typically rural-dominated economy imperfections in financial markets will affect the availability of credit. The total number of banks is B. Each bank b offers deposits D_b to households and loans K_b to potential start-up household enterprises. The solution to the banks' optimisation problem results in a loan-deposit rate spread. This spread (see Gries and Naudé 2010) is to be determined by two factors, namely, the costs of monitoring (c_b) and the concentration of banks measured by the index $\left(1 + \dfrac{1}{B\eta}\right)$. A lower number of banks will increase the concentration of financial intermediaries and widen the interest spread, and similarly if the costs of monitoring increase (decrease) due to for instance a worsening (improvement) of managerial practices, the interest spread will increase (decrease).

Given the loan rate r_1, the expected net present value of the planned household enterprise is

$$EV_m(\tau) = (1-\vartheta)\int_{\tau}^{\infty}\left(p_j - c_{xj}\right)x_j e^{-r_d(t,\tau)(t-\tau)}dt \tag{3.8}$$

where ϑ represent the expected rate of business failure and $(1-\vartheta)$ the expected rate of success. The household will compare these expected profits (or net present value) to wages that can be earned from supplying labour to agriculture (denoted $Ey_{T,i}$) in making its decision to start an enterprise.

If favourable, a new household enterprise will be started up. Because the start-up of a new household enterprise depends on the match between an entrepreneur in the rural sector and opportunities in the city, entrepreneurial abilities, which can be denoted by H, are required to grab a business opportunity, O.

If there is a match, then a household enterprise is established. The number of household enterprises started every period is denoted by γ_N, and over time the growth of household enterprises is $\gamma_N = \dot{N} / N$. This is the speed of urbanisation in the model.

The higher the growth rate γ_N, the more people from rural areas will find employment in cities, and for those that remain in farming, average and marginal productivity will increase, leading to a rise in wage incomes in agriculture. The change in agricultural wages is γ_{y^T}. As per capita incomes (labour productivity) in agriculture grow then rural population will grow as per Eq. (3.2), and the rural sector will expand. For structural transformation and hence development to occur, the speed of urbanisation must exceed the speed of expansion of the rural sector.

3.2.3 Diagrammatic Analysis

The essentials of the model outlined so far can be graphically illustrated with the help of Figs. 3.4, 3.5, and 3.6. Figure 3.4 contains the basic model, depicting the relationship between the central variables outlined above.

The figures have four quadrants. Let us start with Fig. 3.4. In the northeast γ_N-δ quadrant depicts the equilibrium values of γ_N and δ that results from the demand and supply of surplus labour in the city. The downward sloping curve (the *surplus labour curve*) indicates that as the growth rate of city household enterprises γ_N increase, so the surplus labour in the rural areas would decline—hence, the slope is negative.

The upward sloping curve (which is termed the *household enterprise curve*) depicts the converse, namely, that the higher the rate of surplus labour that can be successfully extracted from the rural sector, the more household enterprises can be established in cities. The steady-state equilibrium is at A, which determines the growth rate of the household enterprise sector and the surplus labour rate in the rural sector. At this point, the reallocation of surplus labour to the city is too low (due to the slopes and location of the two curves) so that the growth in the rural area is higher than the growth of the city. Given the population growth rate, an initial equilibrium at point A is consistent with rural stagnation; in other words, the size of the rural area will grow relative to that of the city, as $\gamma_N < \gamma_{y^T}$. Not enough farmers are reallocated via household enterprises to the city. Structurally, the economy remains largely a rural, agriculture-based economy with low productivity, surplus labour in rural areas.

From the discussion above the growth rate of the number of start-ups, i.e. the slope and position of the *household enterprise curve*, is a function of the endogenous surplus labour rate δ_T, and a number of other parameters that have been mentioned in the foregoing, specifically:

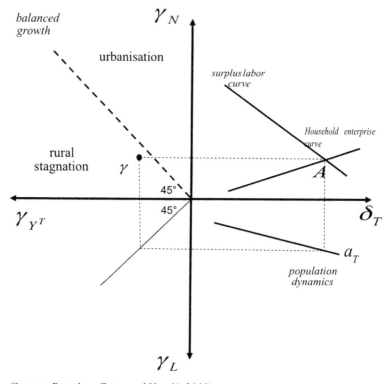

(Source: Based on Gries and Naudé, 2010)

Fig. 3.4 Entrepreneurship and structural transformation: basic model (Source: Based on Gries and Naudé (2010))

- Productivity in agriculture (a_T). If productivity in agriculture improves, then for a given population growth rate γ_L there will be more surplus labour released, which will shift the population dynamics curve in the southeast quadrant inwards and the surplus labour curve outwards, so that there will be growth in the number of household enterprises (see Eq. 3.3).
- Entrepreneurial ability (H). Improvements in H will shift the household enterprise curve upwards, so that there will be growth in the number of household enterprises.
- Finance (B). Improvements in B will shift the household enterprise curve upwards, so that there will be growth in the number of household enterprises.

Improvements in these last two parameters are depicted in Fig. 3.5. If the steady-state equilibrium moves from A to A' (e.g. if entrepreneurial ability improve or access to finance improve) the economy's structure changes from being in rural stagnation (at point γ) to being in the region of 'urbanisation' at point γ'.

The interdependence between entrepreneurship, urbanisation and development in this model is evident from the fact that it can be seen that urbanisation has resulted

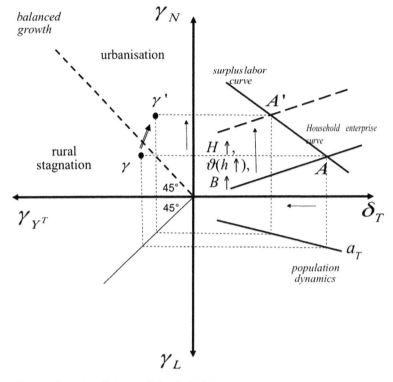

(Source: Based on Gries and Naudé, 2010)

Fig. 3.5 Entrepreneurship and structural transformation: improving entrepreneurial ability and finance (Source: Based on Gries and Naudé (2010))

in an increase in productivity growth and hence per capita GDP, two empirical facts of urbanisation.

Finally, for present purposes, although this is not formally derived here, the slope of the household enterprise curve will depend on the ease with which for a given population growth rate, surplus labour can migrate to the city. One can show that if surplus labour moves to a secondary city, it may be easier given the proximity, ease of access to opportunities and other reasons[5] (see Christiaensen and Todo 2014). Hence, by having more of such agglomerations, the slope of the household enterprise curve can become steeper and urbanisation can occur.

In Fig. 3.6, I show what can happen if it is easier to migrate, because there is a secondary city/rural agglomeration that can absorb farmers. The household enterprise curve swivels (i.e. changes slope) leading to the growth rate in household

---

[5] Christiaensen and Todo (2014) found that migration of labour from rural areas into secondary towns result in greater poverty reduction than migration to capital cities. They find that 'off-farm jobs generated in nearby villages or rural towns may be more readily accessible to the poor given lower thresholds to migrate and better compatibility with their skill set' (p.43).

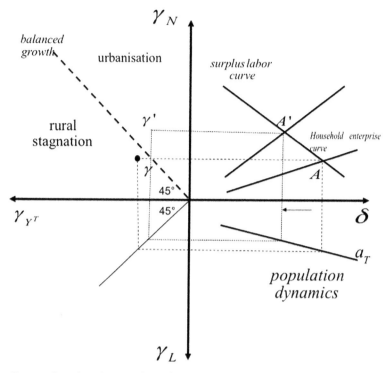

(Source: Based on Gries and Naudé, 2010)

Fig. 3.6 Entrepreneurship and structural transformation: secondary cities (Source: Based on Gries and Naudé (2010))

enterprises exceeding the growth rate in rural incomes and the population share of the city relative to agriculture increasing. It emphasises the importance of migration for urbanisation and development (Naudé et al. 2017) and moreover for the individual welfare of the migrant (Gibson and McKenzie 2012).

In Figs. 3.4 and 3.5, the urbanisation of the economy takes place and is driven by the entrepreneurship. To the extent that urban growth is driven by household enterprise start-ups that supply intermediate services to large firms in the modern sector (who in turn produce for final demand), the service sector assumes an increasingly important role in the economy. This is consistent with the 'stylised facts' of structural change.

Although in this model the productivity of each intermediate service remains constant, total factor productivity increases in the final goods producing sector (manufacturing) the steeper the entrepreneurial start-up curve or the higher it shifts. This can account for a further 'stylised fact' of modern structural change, namely, the higher productivity of the manufacturing sector relative to the services sector (and the rural economy).

In conclusion, in this simple model, for urbanisation to start, the vehicle is entrepreneurial innovation and start-ups. Factors to facilitate this, such as a conducive business environment (e.g. for finance), education and skills formation (e.g. for entrepreneurial ability) has been shown to be essential in the process of urbanisation and hence of development.

3.3   Early Development and Cities

At early stages in a country's development, when most people are living in rural areas (corresponding to point γ in Fig. 3.4), people will start to flock to an emerging city with the hope to find a job (and better health and education) and to start-up a new enterprise by providing a unique good or service (as described in the previous section). Many will succeed: the probability of a job is higher in most countries in a city than in a rural area. But many will not and will resort to survival or necessity self-employment, as opposed to providing the innovative unique goods and services as in the previous section that drives urban development.

Therefore, in reality in cities in developing countries, we find almost without exception large numbers of small, micro informal and formal enterprises providing all kinds of undifferentiated consumer products and services. A lot of these play a role in sustaining the 'shadow economy', including the illegal drug trade, prostitution, people smuggling and the like. It starkly reminds us that entrepreneurship is not always good for development: entrepreneurs can be unproductive and even destructive (Baumol 1990). And the shadow economy never disappears: as cities grow and countries become more urbanised, the size of the shadow economy declines, but it remains a feature to be dealt with (see Fig. 3.7).

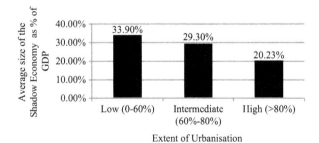

(Source: author's compilation based on data from the World Development Indicators, World Bank and from Manes et al., 2016)

Fig. 3.7 The size of the shadow economy is on average largest in countries that are less urbanised (Source: author's compilation based on data from the World Development Indicators, World Bank and from Manes et al. (2016))

Hence, the policy quest is to design an urban entrepreneurial ecosystem[6] that will offer strong incentives for high-growth, productive and legal entrepreneurship and discourage illegal activities. One of the most import requirements for such an emerging entrepreneurial ecosystem is infrastructure. Infrastructure provides connectivity and energy,[7] which is needed by entrepreneurs for obtaining 'scale and specialisation' (Collier and Venables 2016:394). The difficulty is that sound political decision-making and governance institutions, including land rights, are required for such an ecosystem: in most rapidly developing and urbanising countries these are largely missing. Good urban planning and management skills, including urban policing and dealing with land disputes, may be amongst the most sorely needed in the emerging world today.

Countries at relatively earlier stages of development and urbanisation will, apart from thousands of self-employed entrepreneurs and their small businesses, also attract larger foreign businesses into their growing cities. These foreign 'entrepreneurs' arrive to make use of the pool of low-wage labour, to access raw minerals and natural resources in the hinterland and to sell consumption goods to the local population. As with the case of small and informal businesses, these large foreign businesses can have negative side effects on development. Rent-seeking, lobbying, bribing and exploitation, these are just some of the nefarious activities that large foreign businesses and foreign entrepreneurs have been documented to indulge in the fledgling capitals of developing countries (see, e.g. Burgis 2016). For many developing country cities, this exacerbates the impact of the domination by multinational corporations during the period of colonialism (see, for instance, the account by Jones (2010)).

3.4   Entrepreneurial Hotspots

Once a country's state of development and urbanisation has reached point γ' (see Fig. 3.5) and if these new urban agglomerations are managed well to limit the negative spillovers from entrepreneurship as mentioned, many of the small start-ups will scale up and employ more people. The larger and denser population matters for learning, for sharing ideas and for innovation and technological take-up. Population growth and the density that urbanisation provides to a larger population are essential elements in long-run development. As Derex et al. (2013) put it 'complex cultural traditions—from making fishing nets to tying knots—last longer and improve faster at the hands of larger, more sociable groups. This helps to explain why some groups, such as Tasmanian aboriginals, lost many valuable skills and technologies as their populations shrank'.

---

[6] For a discussion on the concept of an 'entrepreneurial ecosystem' see Isenberg (2010) and Stam and Spigel (2016).

[7] Energy provision is more amenable to being provided by private entrepreneurs as opposed to connectivity, which requires government investment, regulation and coordination (Collier and Venables 2016).

Into these 'larger, more sociable groups' in cities also more foreign entrepreneurs will be attracted. International migration is an essential process to keep fledgling cities vibrant. Think of how cities such as London, New York and Shanghai were built by waves of immigrants.[8] Empirical evidence suggests that culturally diverse cities are more productive and more innovative and better hotspots for entrepreneurship. For instance, Ottaviano and Peri (2006:9) report evidence from the USA that 'US-born citizens living in metropolitan areas where the share of foreign-born increased between 1970 and 1990, experienced a significant increase in their wage and in the rental price of their housing'.

The growing city will become an entrepreneurial hotspot, as local and foreign firms vie to build houses, office spaces, retail centres and public infrastructure. Entrepreneurial support services, such as banks, transport providers, export and import agents, ICT specialists, job placement agencies and business schools will thrive. Urban governments will spend resources on locality marketing and promoting entrepreneurship and small businesses within their jurisdictions, often becoming *entrepreneurial governments* in the process.[9]

Venture capital follows and leads these entrepreneurial hotspots and tend to be concentrated in certain 'global start-up cities': for instance, the volume of venture capital lending in cities like San Francisco and Beijing exceeds the volume of venture capital in an entire country such as Germany (Florida and King 2016).

As a result of the above, productivity, and hence average wages, would rise. Collier and Venables (2016:396) report that empirical evidence suggests that 'productivity in a city of 5m is between 12% and 26% higher than in a city of half a million'. A large volume of empirical research furthermore confirm that formal, urban-based firms are more productive than nonurban and informal firms (see, e.g. Rijkers et al. (2010), Söderbom and Teal (2004) and Van Biesebroeck (2005)) and that wages are higher in cities than elsewhere, on average (Glaeser 1998). If productivity growth is higher in some industries than others in particular city, as a result of that industry reaping economies of scale, we will see some cities starting to specialise, for instance, London or Frankfurt where the financial industry is concentrated or Los Angeles in cultural industries and Bangalore in ICT (Duranton and Puga 2000; Henderson 2010).

Eventually, rising property prices and rents, urban congestion and fierce business competition will mean that the costs of urbanisation will rise. Henderson (2010) reports that 'moving from a city of 250,000 to one of 2.5 million is associated empirically with a 80% increase in commuting times and housing rental prices' (p.519). With property investments becoming profitable, entrepreneurs in the property development sector could start to claim and develop more land for urban expansion. This

---

[8] In the case of the USA, for example, Hirschman and Mogford (2009) document that by 1900 no less than 75% of the populations of the largest cities, including New York, Chicago, Boston and San Francisco consisted of first and second generation immigrants.

[9] Urban governments' initiatives to develop the economies of their cities have been described as being entrepreneurial if these 'involve the creation of new institutions of governance, new links between existing institutions or actors or new ways of exploiting opportunities' (Olsson et al. 2015:137).

property development role of entrepreneurs is important for the continued development of cities and in turn requires adequate land rights and land markets (Collier and Venables 2016). Cities will become larger and or multiply in number, and the spatial form of cities will change shape. Urban sprawl may become a problem. In general, industry will, under pressures of competition, become more technologically advanced.

Entrepreneurial innovation will become an important driver of productivity and economic growth and will be almost exclusively found in larger cities. Jobs will move out of mechanised and automated industry into services—and the high tech, high skill requirements on these services will spurn many new entrepreneurial opportunities. Think of new business models and new markets being created by the intersection of new technologies and entrepreneurship in congested cities, such as *Uber* (solving transport problems) or *AirBnB* (solving short-term accommodation problems) and many other service apps. 'Smart cities' defined as cities that 'uses digital technologies to improve its performance management and well-being, reduces costs and resource consumption and engages more efficiently and effectively with citizens', becomes a trend at this stage (Bezerra et al. 2015:1).

3.5   Quality of Life: The Quest for Sustainability

As the virtuous development cycle of urbanisation and entrepreneurship continues, most people end up living in or near a city. The high concentration of billions of people in a few geographic spots, many of them on the coast, raises the vulnerability of a highly interconnected system of production and consumption to, for instance, natural disasters (e.g. the 2011 Fukushima nuclear disaster caused by a tsunami) or urban terrorism (e.g. the 9/11 attacks on New York). Moreover the rise of cities inevitably has led to concerns about the environmental impact of development, in terms of, for instance, CO_2 emissions and pollution. Figure 3.8 shows the strong positive correlation between the degree of urbanisation and the greenhouse gas emissions (GHG) (e.g. CO_2 emissions) per person in a country.

Fig. 3.8 Urbanisation is associated with increased GHG emissions (Source: author's compilation based on data from the World Development Indicators, World Bank)

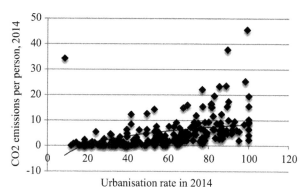

(Source: author's compilation based on data from the World Development Indicators, World Bank)

These risks and dangers pose new challenges for urban management and city planning, as well as create opportunities for entrepreneurs to generate and implement technologies arise to deal with these vulnerabilities and risks. Examples are the development and use of better surveillance systems and prediction models based on 'big data', or the emergence of urban agricultural systems (e.g. vertical farming) aimed at reducing food insecurity and GHG emissions.

At this stage, cities may compete to attract what Richard Florida called the 'creative class' (Florida 2002). Quality of life becomes central. Security, health, education, a beautiful environment and efficient public services, including arts and cultural amenities, are required to attract and keep the 'creative class'. This becomes more acute as declining transport costs make production less place dependent and more human capital dependent (Glaeser 2003).

Around this stage, many cities also become the centres of the circular economy.[10] An example is the Dutch city of Almere, where in 2016 Europe's first 'eco-village' was started that is a completely self-sustainable urban community. An entrepreneur, who plans to roll out the concept across Europe in coming years, drives this initiative.[11] Sustainability, in the city context, is however a wicked problem, and as explained by Woolthuis et al. (2013:94), 'the wickedness lies not only in the technical sophistication, multiple scales and multi-actor character of the problem, it also relates to the fact that there is no consensus on what sustainability is in an urban context'. They call for institutional entrepreneurship[12] to play an important role in finding solutions to environmental challenges in cities by 'creating the conditions in which right solutions can be realized' (Woolthuis et al. 2013:94).

Finally, one can ask what does the future hold, what is the ultimate role of cities, and entrepreneurs, in development? Clearly, with cities becoming more important, the governance of cities and the political power concentrated in cities will make city-states perhaps even more important than the nation state. Will we see growing decentralisation away from the current global governance system of 193 nation states towards a myriad of city-states? Will physical and virtual citizenship of a city-region become more important for one's welfare in the future? There are certainly many scholars and policymakers who think so. According to McKinsey[13] 'the 21st century will not be dominated by America or China, Brazil or India, but by The City. In a world that increasingly appears ungovernable, cities—not states—are the islands of governance on which the future world order will be built'.

Continued technological innovations in ICT and logistics, including the Internet of things, automation, artificial intelligence, and big data will play a key role in driving the further evolution of the city-entrepreneurship nexus. These will by improv-

---

[10] See, for instance, the Circle Cities Programme: https://www.circle-economy.com/tool/cities/#.Wcz4CFzyiS0.

[11] Read more in The Guardian, at: https://www.theguardian.com/sustainable-business/2016/jul/12/eco-village-hi-tech-off-grid-communities-netherlands-circular-housing-regen-effekt.

[12] Institutional entrepreneurship can be defined as 'influential (groups of) individuals or organizations that challenge old, and initiate new, institutions' (Woolthuis et al. 2013:91).

[13] See https://www.mckinsey.com/global-themes/urbanization/when-cities-rule-the-world.

ing government effectiveness and making government and governance even more customer-focused, result in more inclusive societies and proliferation of new business models. For instance, crowd sourcing, 'open government', big data, 'virtual-citizen schemes' (such as Estonia's e-residency[14]) and virtual currencies such as 'Bitcoin' are eroding the traditional nation state and may continue to foster a proliferation of government types and hybrid non-state structures offering 'public services' and 'attracting customers, and deriving revenues without regard to physical territory… [allowing] states to turn public goods into virtual business ventures' (Schnurer 2014, 2015). According to Khanna (2013) 'though most of us might not realize it, "non-state world" describes much of how global society already operates… where growth and innovation have been most successful, a hybrid public-private, domestic-foreign nexus lies beneath the miracle'.

3.6 Summary and Conclusions

The process that was described in the preceding paragraphs described a process of urbanisation and entrepreneurship going together, like a horse and carriage. But like love and marriage, the relationship is neither linear nor certain. There is nothing inevitable in the rise and development of any particular city. Cities do not only generate, they also degenerate: think of many great world cities of the past that have risen and fallen[15] and lost their entrepreneurial splendour—Alexandria, Angkor, Detroit, Glasgow, etc. Cities also differ in economic structure, some cities tend to specialise (like London in finance), while some are more diverse (like Shanghai) depending on its entrepreneurs. For some cities, three million people may be the optimal size, while others seem to be functioning well with 20 million.

There are many questions for researchers and policymakers as to the future of entrepreneurship in cities. Compared to Asia, Africa has few megacities at present: where will these be located in a hundred years from now? Will London decline if the UK exits from the EU? Will the architecturally stunning cities of the Gulf States be sustainable in a post-oil era? How will China's responses to clean up the natural

> **Box 3.1 Entrepreneurship and City Decline: The example of Glasgow**
> The rise and fall of its city of Glasgow illustrates the centrality of entrepreneurship (and luck) in the fortunes of cities. As Frisby (2014) chronicles, Glasgow rose during the eighteenth and nineteenth centuries through

---

[14]The government of Estonia introduced an e-residency scheme in 2014 in terms of which is 'a state-issued secure digital identity for non-residents that allows digital authentication and the digital signing of documents' (see https://e-estonia.com/e-residents/about/).

[15]Reba et al. (2016) has published 'the first spatially explicit dataset of urban settlements from 3700 BC to AD 2000, by digitizing, transcribing, and geocoding historical, archaeological, and census-based urban population data'. See: http://www.nature.com/articles/sdata201634.

entrepreneurs sizing on its favourable location as a harbour and seafaring hub (exploiting the trade winds) and the inventions of the industrial revolution, to become by 1900 the second city of the British Empire. It was considered the best governed city in Europe. It adapted innovatively to many changes in external circumstances: when it lost its position in the tobacco trade after American Independence, it moved on to cotton; when steam ships made its position on the trade winds irrelevant, it became a major producer of ships, producing one fifth of the worlds ships between the 1890 and 1914. But after 1914, its long and slow relative decline set in. By 2014, it had, as reported by Frisby (2014), a 30% unemployment rate, the UK's highest homicide rate and, moreover, the lowest life expectancy in the UK. Its 'entrepreneurship', which helped it to buffer many changes and shocks in the eighteenth and nineteenth centuries, was powerless to prevent its decline. -*Taken from* Naudé (2016b)

environments of its cities influence its role as manufacturer of the world? Will the rise of robotics, networked production and the age of industry 4.0 put a break on the speed of urbanisation that we have seen over the past century? Will entrepreneurs from Silicon Valley (or elsewhere) create their own floating city-states in the Pacific?

The heterogeneity, serendipity and context-specificity that shape the patterns of global urbanisation (no one city is alike) imply that there is much that is still unknown about the specifics of the relationship between cities and its entrepreneurs. After all, urbanisation is a relatively recent phenomenon in human history, considering that in 1800 only 3% of the world's population was urbanised. The most interesting period in the world's urbanisation is clearly still to come.

References

Baumol, W. J. (1990). Entrepreneurship: Productive, unproductive and destructive. *Journal of Political Economy, 98*, 893–921.

Bezerra, R., Nascimento, F. M. S., & Martins, J. S. B. (2015). On computational infrastructure requirements to smart and autonomic cities framework. *2015 IEEE First International Smart Cities Conference (ISC2)*, 25–28 Oct 2015, Guadalajara

Burgis, T. (2016). *The looting machine: Warlords, tycoons, smugglers and the systematic theft of Africa's wealth*. London: William Collins.

Christiaensen, L., & Todo, Y. (2014). Poverty reduction during the rural-urban transformation – The role of the missing middle. *World Development, 63*, 43–58.

Ciccone, A., & Matsuyama, K. (1996). Start-up costs and pecuniary externalities as barriers to economic development. *Journal of Development Economics, 4*, 33–59.

Collier, P., & Venables, A. J. (2016). Urban infrastructure for development. *Oxford Review of Economic Policy, 32*(2), 391–409.

Derex, M., Beugin, M.-P., Godelle, B., & Raymond, M. (2013). Experimental evidence for the influence of group size on cultural complexity. *Nature, 503*, 389–391.

Duranton, G., & Puga, D. (2000). Diversity and specialization in cities: Why and where and when does it matter? *Urban Studies, 37*(3), 533–555.

Florida, R. L. (2002). *The rise of the creative class*. Basic Books.

Florida, R. L., & King, K. M. (2016). *Rise of the global startup city*. Martin Prosperity Institute. Toronto.

Frisby, D. (2014). Glasgow: The rise and fall of a start-up hub, Virgin Entrepreneur, 1 August 2014, at: http://www.virgin.com/entrepreneur/glasgow-the-rise-and-fall-of-a-start-up-hub

Gibson, J., & McKenzie, D. (2012). The economic consequences of 'brain drain' of the best and brightest: Microeconomic evidence from five countries. *Economic Journal, 122*, 339–375.

Glaeser, E. L. (1998). Are cities dying? *Journal of Economic Perspectives, 12*(2), 139–160.

Glaeser, E. L. (2003). The new economics of urban and regional growth. In G. Clark, M. Feldman, & M. Gertler (Eds.), *The Oxford handbook of economic geography* (pp. 83–98). Oxford: Oxford University Press.

Glaeser, E. L. (2007). Entrepreneurship and the City'. *NBER Working Paper No. 13551*. National Bureau for Economic Research. Cambridge, MA.

Glaeser, E. L. (2011). *Triumph of the City: How our greatest invention makes us richer, smarter, greener, healthier and happier*. London: McMillan.

Gries, T., & Naudé, W. (2010). Entrepreneurship and structural economic transformation. *Small Business Economics Journal, 34*(1), 13–29.

Henderson, J. V. (2010). Cities and development. *Journal of Regional Science, 50*(1), 515–540.

Hirschman, C., & Mogford, E. (2009). Immigration and the American industrial revolution from 1880 to 1920. *Social Science Research, 38*(4), 897–920.

Isenberg, D. (2010). The big idea: How to start an entrepreneurial revolution. *Harvard Business Review*. June, pp.1–11.

Jones, G. (2010). Multinational strategies and developing countries in historical perspective, *Working Paper 10–076*, Harvard Business School.

Khanna, P. (2013). The end of the nation-state?. The New York Times, 12 October

Lewis, W. A. (1954). Economic development with unlimited supplies of labour. *The Manchester School, 22*(2), 139–191.

Manes, E., Schneider, F., & Tchetchik, A. (2016). On the boundaries of the shadow economy: An empirical investigation. *IZA Discussion Paper no. 10067*. Bonn: Institute for the Study of Labor

Naudé, W. (2016a). Entrepreneurship and the reallocation of African farmers. *Agrekon, 55*(1), 1–33.

Naudé, W. (2016b). Is European entrepreneurship in crisis?' *IZA Discussion Paper 9817*. Bonn: Institute for the Study of Labor

Naudé, W., Marchand, K., & Siegel, M. (2017). Migration, entrepreneurship and development: Critical questions. *IZA Journal of Migration, 6*, 5.

Olsson, A. R., Westlund, H., & Larsson, J. P. (2015). Entrepreneurial governance for local growth. In K. Kourtit, P. Nijkamp, & R. Stough (Eds.), *The rise of the city: Spatial dynamics in the urban century* (pp. 135–159). Cheltenham: Edward Elgar.

Ottaviano, G. I. P., & Peri, G. (2006). The economic value of cultural diversity: Evidence from US cities. *Journal of Economic Geography, 6*(1), 9–44.

Ovanessoff, A., & Purdy, M. (2011). Global competition 2021: Key capabilities for emerging opportunities. *Strategy & Leadership, 39*(5), 46–55.

Reba, M., Reitsma, F., & Seto, K. C. (2016). Spatializing 6,000 years of global urbanization from 3700 BC to AD 2000. *Science Data, 3*, 160034. https://doi.org/10.1038/sdata.2016.34

Rijkers, B., Söderbom, M., & Loening, J. L. (2010). A rural-urban comparison of manufacturing Enterprise performance in Ethiopia. *World Development, 38*(9), 1278–1296.

Schnurer, E. (2014). Welcome to E-Stonia. *US News,* 4 December.

Schnurer, E. (2015). E-Stonia and the future of the cyberstate. *Foreign Affairs*, (28 January).

Söderbom, M., & Teal, F. (2004). Size and efficiency in African manufacturing firms: Evidence from firm-level panel data. *Journal of Development Economics, 73*(1), 369–394.

Stam, E. & Spigel, B. (2016). Entrepreneurial ecosystems', *Discussion Paper Series 16–13*, Tjallings C. Koopmans Research Institute.

Van Biesebroeck, J. (2005). Exporting raises productivity in sub-Saharan manufacturing firms. *Journal of International Economics, 67*(2), 373–391.
Woolthuis, R. K., Hooimeijer, F., Bossink, B., Mulder, G., & Brouwer, J. (2013). Institutional entrepreneurship in sustainable urban development: Dutch successes as inspiration for transformation. *Journal of Cleaner Production, 50*, 91–100.
Zhang, J. (2002). Urbanization, population transition, and growth. *Oxford Economic Papers, 54*, 91–117.

Chapter 4
Social Sustainability, Housing and Alienation

Jamileh Manoochehri

4.1 Introduction

> The economic engine that is capital circulation and accumulation gobbles up whole cities only to spit out new urban forms in spite of the resistance of the people who feel alienated entirely from the processes that not only reshape the environment in which they live but also redefine the kind of person they must become to survive (Gorz 1989, p.47).

The urban environment and with it housing are the sites of social and individual life. Significant changes to these have great impact on people's lives. Housing, like more and more aspects of our lives, is monetized and defined as a commodity. As access to housing as a home and shelter is conducted via the market, it moves out of reach of those on low income who do not qualify under stringent requirements for access to the diminishing stock available to local authorities.

In the UK, state policies since the introduction of the Right-to-Buy policy in 1980[1] and the promotion of home ownership as a pillar of a 'property-owning democracy'[2] have converted what used to be a home (rented or owned) into real estate. This monetization is not limited to housing and applies to most if not all aspects of life in post-industrial capitalist states[3]. In the workplace we sell our labour for wages that lose their purchasing power all the time, as the pace of work is

---

[1] The Right to Buy was proposed in the Conservative manifesto of 1979 and was enshrined in the Housing Act 1980 in England and Wales and the Housing Tenants Rights, etc. [Scotland] Act 1980.

[2] Margaret Thatcher's first speech as Conservative Party Leader, given in Blackpool, 1975 [Available from: http://www.britishpoliticalspeech.org/speech-archive.htm?speech=121, Accessed: 01.08.2017].

[3] Auction of Presidio Terrace in San Francisco in August 2017; The Auction of AAMBY Valley City in the District of Pune in Maharashtra, India, advertised in Time Magazine, August 2017; Private Notice pinned to Tree in Street in Tehran offering to sell body organs.

J. Manoochehri (✉)
Leicester School of Architecture, De Montfort University, The Gateway, Leicester LE1 9BH
De Montfort University, Leicester, UK
e-mail: jmanoochehri@dmu.ac.uk

© Springer International Publishing AG, part of Springer Nature 2018
M. Dastbaz et al. (eds.), *Smart Futures, Challenges of Urbanisation, and Social Sustainability*, https://doi.org/10.1007/978-3-319-74549-7_4

intensified and means of subsistence become more costly. The means of access to what we need is the market; the only way we can purchase any commodity is by earning money through selling our labour. This process is the key to how our relations with almost everything around us is defined.

Charlie Chaplin's Modern Times (movie released in 1936) visualizes, with great effect, the way the mechanization of production disempowers the worker, turning him into an extension of the machine and part of its process, rhythm and logic. What Chaplin satirizes is part and parcel of an alienation, where the product of labour embodied in an object leads to the objectification of labour. This objectification does not stop at the factory gates; it enters all aspects of the worker's life. At the core of the objectification is the monetization of labour. It may be argued that the extension of this leads us to see the same form of alienation where the exchange value of things surpasses their use value in importance, making them devoid of their substance and finding meaning only in the market. The fact that the worth of every thing and aspect of our lives is gauged through money has changed the way we relate to them.

This chapter argues that this process is an integral part of alienation and is manifested in how we relate to others and how we define our own worth. The mediation of money affects how we relate to the built environment and our dwelling too, viewing them less as use value – what function they perform – and more as exchange value – what they are worth in the market – or what we have to pay to be allowed to partake of them. In the argument around sustainability, the economic aspect of sustainability links into this in a particular way. Arguments for energy efficiency and use of advanced technology tend to gauge success according to monetary savings. Here, it is argued that the effects of alienation are detrimental to the way the home and housing are viewed and that urban planners, architects and designers need to consider this in order to design sustainable cities and homes – socially, environmentally and economically.

4.2   Sustainability Agenda: At Whose Service?

The nature of the interest in sustainability as a value system is worth reflecting on; is it a philosophy or ideology? When considering the social sustainability agenda, the policy documents of the United Nations suggest a set of values that have an ameliorating function. As Loukola and Kyllonen (2005) suggest, the 'nonfinancial considerations' associated with sustainability have entered 'the moral and political debate' as a counterpoint to the 'economizing' of society's functions and goals. In their study of the philosophies of sustainability, they propose that the approaches towards sustainability emanate from anthropocentrism or ecocentrism and are based on three different conceptions of sustainability – 'critical natural capital' (*necessary or critical to the production and reproduction of human well-being)*, 'irreversible nature' (*indispensable aspects of nature whose loss would be irreversible*) and 'the value of nature' (*indispensable aspect of nature that has an intrinsic value as well*).

In recent years, the concept of ecology has been favoured over sustainability. There is a distinction made between sustainability that is sought for the sake of nature itself (ecocentric) and that based on societal issues (anthropocentric). These two approaches are poised against each other (Harvey 2002) based on a contradiction generated by human versus nature relations. However, there is much to argue that these are interconnected and not mutually exclusive. Social sustainability is seen as an excessively general and idealistic notion (Loukola and Kyllonen 2005). However, it is conceded that there are clear characteristics to it such as 'determining social goals and making distributional decisions' (ibid). The sustainability agenda has gained both social and temporal aspects and has been broadened to include, as recipients of distribution, future human generations and significantly, to some extent, also 'non-human objects' (ibid, pp.7–8). On the issue of monetizing the world's resources, Loukola and Kyllonen reflect on the fact that estimating a market price for the aspects of nature is a 'challenging task, if not altogether impossible' and that no economic method has been developed 'that would transform all the values people attach to natural objects to exchange values priced in the market' (ibid, p.7). This bid to remove the agenda from the monetization that is predominant is an important one, in the way (social) sustainability is perceived and promoted.

There are arguments against the sustainability agenda based on the view that it curtails development by countries and people with the smallest carbon footprint or alternatively that the effects of climate change have not been proven; therefore, the exploitation of natural resources should not be impeded, for example, the construction of the disputed pipeline in Alaska, to the detriment of the environment and the local people with a stake in that environment.

Harvey (2002) takes a critical view of the sustainability agenda and argues that humans are 'active subjects' who have 'accumulated massive powers to transform the world' in the 'evolutionary game'. How we exercise these powers defines what we will become. This is the focus of debate 'among capitalists and their allies (many of whom are obsessed with the issue of long-term sustainability) as among those who seek alternatives' (ibid). Harvey's stance on the subject is clearly an anthropocentric one, maintaining that human beings 'cannot ever avoid … asserting our own species identity, being expressive of who we are and what we become, and putting our species capacities and powers to work in the world we inhabit' (Harvey 2002, p.213). Harvey's assessment of capitalism's need and wish to sustain the status quo and balance of power in its favour is irrefutable. There are many examples of the warnings of an environmental crisis having helped legitimize policies that have been fundamentally regressive and oppressive, such as the example he cites of the Conservative Party's political decision on closing the coal mines in the UK in the 1980s under the cloak of freeing the country's energy industry from its dependence on coal. However, there is a clearly progressive aspect to the argument for a sustainability agenda. This is supported primarily by the fact that as capital maximizes its profits, the excessive consumption of worldwide resources benefits it at the expense of large populations who are impoverished and forced to lead an unfulfilled and miserable existence. In this sense, Harvey's point is valid that 'the distinction between the production/prevention of risks and the capitalistic bias towards consumption/commodification of cures has significance' (Harvey 2002, p.221). In the conflict of interest between the rich and the poor, the sustainability argument can and must be used in favour of the disenfranchised, and the poor who are 'located' and 'dislocated' by policymakers, dispersed and made homeless through a variety of means – from selling off state assets in order to withdraw the state from the provision of social services and goods, to cynical proxy wars maintained across the globe in order to serve the interests of capital and leading to impoverishment and mass displacement of millions. The sustainability agenda has the potential to be articulated constructively and to become the site of the dialectical change, a part of which the contradiction of human and nature is thought to be. Exploitative development is harmful to humans as well as nature. We appear to have reached a stage where we cannot view the nature-human relations as oppressive to humans. 'Man' has 'conquered' nature to such an extent that his continued exploitation of it cannot be justified easily, although the duality of man-nature and the concept of nature as oppressive and as one to be tamed and conquered, which dominated during the Enlightenment until today, persists. Enlightenment thinkers aimed 'to develop objective science, universal morality and law, and autonomous art according to their inner logic' (Habermas 1983, p.9, cited in Harvey 1990), all based on the scientific domination of nature. The control of natural forces was part of the effort for the happiness of humankind. This view of nature as 'other' persists despite the scientific exploitation of nature and the living world having been unrelenting for at least the last two centuries. Exploitative relations that emanate from social power structures are attributed to nature in some feminist texts. In Cubonik's *Xenofeminism*, 'we need new affordances of perception and action unblinkered by naturalized identities', and '"nature" shall no longer be a refuge of injustice' (Cuboniks 2016). However, sustainability is not too far removed from the aim of respecting people's rights to maintain their local, communal, social, familial and individual connections with all the contradictions and opportunities that they may entail. Communities and social relations contain support systems as well as potential for oppression. However, in the context of the urban environment and housing, the resolution of these contradictory relations is not helped when families or communities are forced to disperse. Such dispersals are bound to disrupt the process of dialectical change that may occur through the interactions of different actors in the conflict. It is pertinent to evaluate the effect of commodification and private property accumulation on our relations with our environment and our homes and how the sustainability argument must address the adverse effects of development on people.

Seeing social sustainability from this point of view requires that the threats to social and individual well-being are identified. As mentioned earlier, at the core of the social and economic relations in advanced capitalist countries is the necessity for the accumulation of wealth, itself based on the mechanisms of exchange and the exchange value of all products, and ultimately all things. The commodification of all things affects societal values and relations. Inevitably, this impacts on the individuals within the society and becomes visible in the way we perceive ourselves, our relationships with others and our environment and homes.

4.3   Sustainability, Social Policies and Housing

Social sustainability is bound to be connected to the anthropocentric conception of sustainability. The social aspects of the environment play out in the society and social space in general but have relevance in the home too. The city is shaped by the requirements of the capitalist state, and social policy is geared, by and large, towards the reproduction of capitalism and its means of production. Left to its own devices, this and accumulation of capital become the main focus of policy-making activities of the state. The accumulation of capital requires the individualization of those who provide the labour that generates the profit accrued by capital investments. The living condition of the labour force or potential labour force connects this issue to social housing and the policies that relate to the housing of the low income.

In policy documents, among the most pervasive concerns about social sustainability are factors that emanate from poverty, even though they are not given the prominence that they might. The fact that social sustainability is recognized as a pillar of the sustainability agenda is evidence of the relevance of the idea of society and the 'social' and the potential they have of being constructive means of identifying our collective needs and desires. In his contribution to the UN report, Brundtland[4] identified sustainable development with a number of characteristics – a development that '*meets the needs of the present without compromising the ability of future generations to meet their own needs*'. In this definition we encounter the concept of 'needs' especially '*the essential needs of the world's poor*' and the idea of '*limitations imposed by the state of technology and social organization on the environment's ability to meet present and future needs*'. 'Development' which has been taken to be the key to prosperity of national states has been defined here. Sustainable development requires the '*progressive transformation of economy and society*' and '*changes to access to resources and in the distribution of costs and benefits*' are needed in order to work towards '*social equity*'. The satisfaction of '*human needs and aspirations*' is seen as '*the major objective of development*'. These definitions are rooted in an understanding of the social aspect of sustainable development, as it is based on meeting the basic needs of everyone and allowing for the opportunity for a better life for all. In a world that is increasingly linked up and each part affects another, overconsumption and indulgence by some have a detrimental effect on other human beings as well as the environment and the earth. As Brundtland states, 'living standards that go beyond the basic minimum are sustainable only if consumption standards everywhere have regard for long-term sustainability. Yet many of us live beyond the world's ecological means, for instance in our patterns of energy use' (ibid).

In the UK policy documents, the social aspect of sustainability is covered under the heading: society. In the Department of Communities and Local Government papers and policies, sustainable development is measured by indicators that cover:

---

[4] [Gro Harlem Brundtland, Oslo, 20 March 1987, Report of the World Commission on Environment and Development: Our Common Future, Chapter 2, Conclusion, section I].

economy, society and environment.[5] Those relating to society include, among others, social capital and housing provision.[6] There are factors that have massive social impact but are considered under economy, namely, economic prosperity, long-term unemployment, poverty, knowledge and skills, population demographics and debt. Clearly, the existence and exacerbation of long-term unemployment and poverty militate against a sustainable society – one that can subsist and prosper. The concept of 'social equity' is notable for its absence here.

The United Nations' 17 Sustainable Development Goals[7] (UN 2017) set a clearly ethical and social agenda. The report's foreword holds that sustainable development 'depends fundamentally on upholding human rights and ensuring peace and security' (UN 2017, p.2). It has an element of social justice, calling for an agenda that would reduce inequalities and reach those most at risk. It also declares its resolve to work towards preventing conflict and sustaining peace (ibid). These objectives, albeit functioning within the capitalist system, are bound to have a positive impact on the lives of the poor.

Statistics on development portray massive disparity across the world. From 2000 to 2015, 'expansion of urban land outpaced the growth of urban populations, resulting in urban sprawl', and while the proportion of urban populations living in slums declined by 20% since 2000, their numbers still continue to grow (ibid, p.40); 'the material footprint per capita (amount of raw materials extracted globally that are used to meet the domestic final consumption demand of a country) rose from 48.5 billion metric tons in 2000 to 69.3 billion metric tons in 2010'.[8] The statistics also portray a sharp inequality in the share of wealth and the climate impact across the globe. While Australia and New Zealand had the highest material footprint per capita of 34.7 metric tons per person, the sub-Saharan Africa had the lowest at 2.5 metric tons per person. The report confirms that 'much of the raw material extracted globally goes to serve the consumption needs and habits of individuals in developed regions' (UN 2017, p.42). This consumption, promoted incessantly to maintain the profits of capital, together with the increasing monetization and commodification of all things, is a threat both in ecocentrist and anthropocentrist terms.

---

[5] 'Sustainable Development Indicators' published 18th July 2013 by Defra.
   https://www.gov.uk/government/publications/sustainable-development-indicators-sdis, [Accessed: 15.08.2017].

[6] The 11 indicators are healthy life expectancy, social capital, social mobility in adulthood, housing provision, avoidable mortality, obesity, lifestyles, infant health, air quality, noise and fuel poverty.

[7] Sustainable Development Goals (UN 2017).
   [Available: https://unstats.un.org/sdgs/files/report/2017/TheSustainableDevelopmentGoals Report2017.pdf, Accessed: 16.08.2017].

[8] Eastern and South-eastern Asia – mainly China (28.6 billion metric tons) and Europe and Northern America (21.9 billion metric tons) (UN 2017, p.42).

4.4 Housing: Use Value Versus Exchange Value

The context of private ownership and the switch to a semi-commodified status through the welfare state and a burgeoning social housing sector changed the landscape of housing throughout the twentieth century in the UK. Whereas 23% of households in 1918 were privately owned and 76% privately rented, in the 2014–2015 (Office for National Statistics figures – ONS) private ownership [had] risen to 63/6%, social renting [was] at 17.4% and private renting at 19%. Viewed in the broadest sense and taking account of mortgages and outright ownership, the housing stock in England and Wales was split by 1/3 being owner-occupied, 1/3 owner-occupied and part-owned by banks and less than 1/3 social rented. The main change has been the 1/3 share for the banks (ONS, Table FT1101 (S101): Trends in tenure, 2015).[9]

The right to shelter is not open to dispute internationally. Article 25 of the Universal Declaration of Human Rights recognizes the right to shelter in its definition of an adequate standard of living: 'Everyone has the right to a standard of living adequate for the health and well-being of himself and of his family, including food, clothing, housing and medical care and necessary social services, and the right to security in the event of unemployment, sickness, disability, widowhood, old age' (UN 2016). However, for shelter to meet human needs, it needs to satisfy social criteria and those relating to well-being, with all its individual and social dimensions.

The debate about housing and its crisis invariably centres on its availability and its affordability. After this, we have the debate about space standards and in how small a space can a family, a couple or an individual survive – for living and prospering are no longer the criteria to be met for those who cannot afford market rates. Privatization of social housing in the UK since the 1980s has been critical in creating the current landscape where the market determines the chances of accessing housing. The change of homes into property investment and market commodities has meant that dwellings have acquired an identity other than shelter and home – their use value. The rise in property prices has meant that the monetary value of housing has overshadowed its value as shelter and dwelling. Housing is important to our lives in different ways. The most obvious is housing as physical shelter, but housing and dwelling are also the locus of different scales of social relations – the family being one representation of this, also of self-expression and self-fulfillment.

It can be argued that when housing becomes first and foremost an asset for the purpose of accumulation of value for the purpose of exchange (sale), its character changes. It becomes a part of an alienating form of existence for its inhabitant (and

---

[9] Housing Act 1980, Chapter 51, 'An Act to give security of tenure, and the right to buy their homes, to tenants of local authorities and other bodies; to make other provision with respect to those and other tenants; to amend the law about housing finance in the public sector; to make other provision with respect to housing; to restrict the discretion of the court in making orders for possession of land; and for connected purposes' 8th August 1980 [http://www.legislation.gov.uk/ukpga/1980/51/introduction].

owner). In social housing the dynamics of the market debase the real value of the dwellings. The commodity character of the housing and the land it sits on dissociates it from its principal role and becomes a fundamental part of the disconnection between the material reality of our need for shelter and its role as a market commodity. This aspect of housing opens it to dynamics similar to other commodified aspects of social life, namely, labour. In the capitalist mode of production, labour itself becomes a commodity. Labour produces not only commodities; it produces itself and the worker as a commodity. The product of labour, 'embodied in an object, becomes the objectification of labour'; the realization of labour 'becomes the loss of realization for the workers, objectification becomes the loss of the object and bondage to it', and these dynamics are described as 'appropriation as estrangement, as alienation' (Marx 1844, pp. 28–29). This appropriation is part and parcel of commodification in the field of social housing. When the prerequisite for entering the market is accumulation of disposable income, choice and self-agency through purchase of commodities, peripheral participation in the market only represents a different level of alienation. Is this relationship with the dwelling not an alienated one, disconnected and mediated by the market?

4.5   Alienation

The term alienation is used extensively and in many disciplines: in psychiatry, as a 'state of depersonalization and loss of identity in which the self seems unreal', in law as the transfer of the ownership of property rights and in statistics as 'the lack of correlations in the variation of two measurable variates over a population' (Dictionary 2017). The concept of alienation has been traced from the works of Hegel, Rousseau, Locke, Smith and Feuerbach to Marx (Wendling 2009; Morrison 1998). The term was first used as a philosophic concept in the work of Hegel who, in 1807, used the term 'estrangement' in his work *The Phenomenology of the Mind* in order to 'outline a framework for the development of human consciousness'. In his theory of development, Hegel puts forward the idea that human beings essentially strive to realize themselves in history, a process he referred to as 'self-actualization'. In doing so, individuals would encounter 'oppositions' in which the external world acts to negate the individual by 'shutting out their existence' (Morrison 1998, p.88). Hegel's idea that an individual could 'experience themselves as not fully human' and that the idea of the 'self' could be incomplete fitted the experience of modernism and its association with the fragmentation of human experience (ibid, p.88). Feuerbach paved the way for giving a materialist view of the concept, by criticizing Hegel's position for believing in a 'philosophical world which ruled over the real world', thus duplicating theology. He claimed that philosophy and religion 'both constituted human alienation in that they misrepresented reality and humanity' (ibid, p.89). Marx's critique of Feuerbach, in turn, was that he did not go far enough in stressing the 'material realm over the ideal'. In *The German Ideology*, he rejected Feuerbach's resolving of the 'religious essence into the human essence'. For Marx, 'the human essence is not an abstract [thing], inherent in the single individual. 'In reality it is the ensemble of social relations' (Marx and Engels, 1845, cited in Morrison 1998, p.90).

Auction of AAMBY Valley City, Pune District, India, advertised in Time Magazine, USA

Presidio Terrace – San Francisco sold in Auction

Private Notice, addressed to 'wealthy patients or ill persons who are wealthy', offering body organs for sale

The concept of alienation is a significant one, because it can help understand the conditions that give rise to it and how, if at all possible, it might be countered. As Wendling (2009) suggests, 'in Capital, Marx attempted to reveal those forms of life peculiar to capitalist alienation from within the norms established by this alienation: that is in an environment in which alienation has become so definitive and real that it disappears as a concept and form of analysis'. In such conditions, understanding the concept becomes more difficult but equally necessary.

In his early manuscripts of 1844, Marx explains the conditions in which alienation occurs:

'[The worker's labour] is therefore not the satisfaction of a need but a mere means to satisfy needs outside itself'… 'It belongs to another, it is a loss of his self' (Marx 1844, pp.59–60), and 'The worker becomes all the poorer the more wealth he produces, the more his production increases in power and size. The worker becomes an ever cheaper commodity the more commodities he creates. The devaluation of the world of men is in direct proportion to the increasing value of the world of things. Labour produces not only commodities; it produces itself and the worker as a commodity – and this at the same rate at which it produces commodities in general. This fact expresses merely that the object which labour produces – labour's product – confronts it as something alien, as a power independent of the producer. The product of labour is labour which has been embodied in an object, which has become material: it is the objectification of labour. Labour's realization is its objectification. Under these economic conditions this realization of labour appears as loss of realization for the workers' objectification as loss of the object and bondage to it; appropriation as estrangement, as alienation' (Marx pp. 28–29). Human activity, production and sense of self become connected. This realization is shared by many during the latter part of the nineteenth century. The connection of human well-being and social productivity is visible in the reaction to the early years of capitalist production. William Morris reacted 'against the de-skilling of craft workers through machine and factory production', and his efforts for promoting 'a new artisan culture' were a manifestation of this realization (Harvey 1990, p.23).

One of the important aspects of the idea of alienation as understood by Marx was the idea that 'human beings define themselves in nature and history primarily through their laboring activities'. Inherent in this idea of society is also the idea that 'human beings make society, and at some point society is a natural extension of their nature and their being – it reflects them and they feel at home in it' (Morrison 1998, p. 91).

At the core of these conditions is the issue of commodification of all aspects of our lives, starting with private property. 'Private property, as the material, summary expression of alienated labour, embraces both relations – the relations of the worker to work and to the product of his labour and to the non-worker, and the relation of the non-worker to the worker and to the product of his labour' (Marx 1844, p.35). The resulting alienation is played out not only through the process of industrial production but also, increasingly, in aspects of our everyday life. The concept of alienation develops as does capitalism, hence the way the concept is traced in capital, and Marx's increasing attention to describing alienation, not simply as a feature of capitalist distribution, but as a feature of life in capitalism as a whole (Wendling 2009).

The money economy also changes the bonds and relations that make up 'traditional' communities so that 'money becomes the real community' (Marx cited in Harvey 1990, p.100). Exchange relations dominate our relations, and just as with alienated labour, money appears more as 'a power external to and independent of the producers'. Money and market exchange frame social relationships between things' in 'the fetishism of commodities' (ibid). In *Conditions of Postmodernity*, Harvey (1990) describes a number of characteristics that define postmodernity, and among these are two that, this chapter proposes, are keys to the issue of alienation: the creation of fiction through symbols and fragmentation. In order to function effectively, money must be replaced by symbols such as coins, paper currency and credit, an 'arbitrary fiction'. Through these 'arbitrary fictions', labour and production are represented; however, 'in the absence of social labour, all money would become worthless. But it is only through money that social labour can be represented at all' (Marx cited in Harvey 1990). Also, 'even though money is the signifier of the value of social labour, the perpetual danger looms that the signifier will itself become the object of human greed and of human desire' and '[m]oney becomes the object of desire' and has the capacity to unify due to its ability to 'accommodate individualism, otherness and extraordinary social fragmentation'. Harvey (1990) writes, 'participation in market exchange presupposes a certain division of labour as well as a capacity to separate (alienate) oneself from one's own product. The result is an estrangement from the product of one's own experience, a fragmentation of social tasks and a separation of the subjective meaning of a process of production from the objective market valuation of the product' (ibid, p.103).

Alienation and the effects of the capitalist production are studied in terms of social relations and the city. Foucault's (1972) study of the exertion of power and the prisons, Bourdieu's (1977) study of collective rhythms and temporal forms and spatial structures, De Certeau's (1984) 'pedestrian rhetoric' and Lefebvre's spatial practices in the city (cited in Harvey 1990) study power relations within the social domain. How do these findings affect the home, and by extension, housing? Bachelard writes about the house in detail but does so from a phenomenological perspective. The house is a space of memories but as such becomes a nostalgic view of the place.

On the significance of spatial and temporal practices, Harvey writes, '[T]he history of social change is in part captured by the history of the conceptions of space and time, and the ideological uses to which those conceptions might be put' and that in order to transform society we must deal with the issue of 'the transformation of spatial and temporal conceptions and practices' (Harvey, 1990, p.218). The study of the home cannot be excluded from such a project. Our needs and social relations find their most acute expression in our access to and occupation of our dwellings. As a commodity, does housing become part of the social consumption? Groz (1989) writes, the 'economic rationalization of work' in capitalist development of technological powers, produces 'individuals who, being alienated in their work, will, necessarily be alienated in their consumption as well, and eventually, in their needs' (Groz 1989, cited in Harvey 2015, p.270). However, having such a clear use value, and being an absolute necessity as shelter, the alienation that becomes associated with housing is bound to be different from the 'needs' that the system manufactures in order to increase consumption of more goods. Housing is on a par with food as a basic need.

Alienation is interpreted in many ways, but the common features of its manifestation that are repeated over and again are a sense of deprivation and dispossession, loss and anger. We might recognize alienation in the form of the fissure of local and societal connections – as the disconnection that the worker feels from the product of her labour and finds spatial expression in an inability to mediate in her environment. Might this fissure be said to find expression in her disconnection with her natural networks of support? Just as the early worker was deprived of the chance to give expression to her creative self through making and through her work, owning the product of her labour, she is now unable to affect change in her immediate environment. With this spatial form of alienation, its expression may be found in the absence of self-agency, and self-actualization of some sort, just as the labourer finds meaning in her work and the product of her labour.

If alienation is the absence of the ability and agency to own the product of one's labour, then powerlessness in the environmental conditions is a similar experience to it. In this context does the natural home become not only shelter but also the potential site of enactment and self-actualization? Technological advances have changed our working and social interactions, and to some extent they have altered our relationships with both our homes and the built environment. Whatever the definition or composition of the home, there is an inherent association with the family – it serves to accommodate the unit of human society. Notably, the role of this social unit has, increasingly, transformed over time – and the relationships within the home are continuously subject to change. Technological advances have decentralized the relationship within the family and allow for similar decentralization in the workplace, some of these with benefit in empowering the worker, in one sense, and disempowering in another. We are more flexible in terms of access to information via the Internet, digital means of communication and working from home (in case of some white-collar workers), but our free time is diminishing, and the employers' demand on the workers' time is spreading into their whole time.

The home is not immune from the effects of technological advances and commodity relations. The objectification of labour is extended to finding meaning in owning more and more stuff, mostly useless in the true meaning of the term but important for meeting the manufactured 'need' to drive consumption. The space of the home is packed with so many goods that there is little room for the human occupant. However, we face more problems than compete with our accumulated stuff in order to be comfortable in our homes. The advance of technology, meant to create comfort and reduce energy, is gradually creating greater disconnection with our immediate environment. We introduce increasing layers of detachment from the adjacent environment, the street, the neighbour as well as from other occupants. The individualization that started at the workplace is extended into the home.

If the home is associated with social relations, the family and even relationship between the individual and the workplace – as working from home is becoming more commonplace – then some key components of the home are associated with connections and networks, projection (Cieraad 2015), happiness and contentment. These definitions of the home open up the different aspects of alienation as the condition in which the product of our labour turns into something against us (in the form of commodity), the break in local and societal connections and the inability to mediate in our environment.

The current problem with housing provision in the UK, in great part, relates to the fact that key workers in health, education and other public services are at income levels too low to have access to housing in areas of the city where their labour is required as the property prices are too high. The fact that housing is also an asset has created an additional problem in that the local authority stock in desirable areas with rocketing land prices appears wasteful. Why should local authorities retain land of high market value only to accommodate the unemployed or the low income? The conclusion that tenants of such properties should be moved to where housing of lower market value exists becomes inevitable when every aspect of human life has a price. The dispersal of people in such a way would not be too dissimilar to what we are also experiencing in other aspects of our lives. This dynamic relates to what Harvey calls the 'irrationality of capitalism' that features 'immense human suffering and unmet need' (Harvey 2010, p.215). This unmet need has other effects too – such as co-existence of surplus capital and surplus labour. Taking the conditions in London, the high value of the land and property has put access to desirable or

affordable housing out of the reach of many. This dynamic has led to a form of segregation of communities according to income, and moreover access to housing is
increasingly through the markets. Housing appears to bear a kind of additional tax.
As Edwards (2016) explains besides the tax that is paid to the state and local government for public services, there appears to be another 'unofficial tax paid as rent to
landlords, financial institutions, developers and established owner-occupiers'
(Edwards 2016, p.233). This 'tax' is so unsustainable that local authorities wishing
to make more of their assets start to plan to move people from high-cost areas to
peripheral and low-cost areas of the region or the country. Such policies are accom-
panied by dispersal and the fissure of local and societal connections, as well as the
inability to mediate in our environment. Moreover, housing and the home have a
role additional to shelter and other than as assets. The advancement of technology
and the possibility of virtual connections have meant that the home can be a place
of work and employment too. In 2013 in the UK, those working from home
amounted to 13.7% of all those in employment (Office for National Statistics 2014).

Considering the multiple roles that the home plays in our societal, personal and
professional life, our relationship with it becomes even more significant. The commodification (privatization) of housing and, as Edwards (2016) refers to it, the
'financialization' of access to it create an unwelcome mediator between individuals
and their shelter. The side effect of the Right-to-Buy policy in the UK has been the
emergence of a new social and economic phenomenon – the 'accidental' landlord –
private landlords who buy properties because they have access to sufficient capital
and salaries in order to secure buy-to-let mortgages, and who accumulate 'property
portfolios' for renting, as investment and as a substitute to other forms of financial
investment. The relationship between these landlords and the inhabitants of their
properties is an awkward one, severely distorted by market relations.

4.6   Fragmentation

How do these dynamics of the market, and inability to enter it, affect the non-
participants? In the market-based relationship between capital and labour, as the
latter becomes a commodity, she also becomes individualized and unshackled from
family or community connections. The movement of people once seen in the form
of nannies or nurses and doctors from countries such as the Philippines and India to
wealthy countries in the West, having left their own families behind, is being replicated at a national and regional scale in the destination countries. The break in local
and societal connections is the typical dynamic of this stage of capitalism, and it
brings with it the disjunction between our wish for being whole and the fragmentary
nature of our relations with our environment. In our existence as workers,[10] our
freedom to live as whole human beings is curtailed.

---

[10] Taking the definitions of worker in its broadest sense of functioning in order to earn a living.

Fragmentation, this paper argues, is a part of the prevailing alienation. However, the laws of dialectics assume that change occurs through the interconnections of oppositions, where quantitative changes in these oppositions affect qualitative changes. If the counterbalance of alienation is found in self-agency, as a unifying process, its qualitative change should lead to the change in the balance between the two. The idea of self-agency, as with alienation, itself, is an abstract one, but it can be defined in concrete terms and in terms of its relations with its environment.

Our exercise of self-agency merely within the confines of 'eating, drinking, pro-creating' or at the most in the 'dwelling and in dressing-up etc.', a condition not dissimilar to its critique in 1844 (Marx 1844, p.30), can be understood in terms of alienation. Self-agency taken to mean having the degree of autonomy and ability to make decisions about how we live in our environment and form relationships with the people around us. The conditions of alienation and fragmentation only serve to strengthen the strings that attach labour to capital.

Harvey describes these in terms of 'capitalism circulat[ing] through the body of the laborer (sic.) as variable capital and thereby turn[ing] the laborer into a mere appendage of the circulation of capital itself' (Harvey 1982, p.157). The alienation in this condition permeates and is manifested in multiple aspects of our lives. '... People could be alienated from their work and activities'; ... 'they might be alienated from each other through excessive competitiveness'... 'and they might be alienated from their own essence or human-ness' (Shields 1999, citing Marx 1975).

Alienation has its spatial manifestations. This paper contends that one of these manifestations is in the absence of self-agency, the ability to mitigate the effects of our living environment, the absence of the possibility of enactment and self-actualization, connections and networks, projection[11] of self in the space, happiness and contentment. Alienation has its root in the modes of production but does not end there; it extends to most aspects of our daily life. Lefebvre's idea of 'dis-alienation' as a counterpoint to alienation regards each activity as having the potential for either (cited in Shields 1999). Alienation and disinterest were the by-products of activities that became routine and skills that have been mastered, and the negative quality of alienation was dialectically entwined with dis-alienation and whole-hearted involvement (Shields 1999). Whereas Lefebvre's writings focus much on urban and social space, his categorization of space in his social theory of space specify the physical and cultural aspects of social spatialization. Lefebvre's proposal for a threefold dialectic within spatialization is described by Shields as

---

[11] The term projection is used here in the sense that Cieraad uses it. Cieraard studies the home from an anthropological point of view and draws attention to the concept of 'projections'. This idea resembles the idea of potentialities; however, its relevance here is the way it demonstrates the importance we place on our self-agency and links outside of ourselves, not only within temporal immediacy but with a view to our future existence. Cieraard's study found that the 'temporal restructuring of home in a past, present, and future home, and the intertwining of memories of past homes and projections of future homes' was important to the individual at a psychological level. However this was not limited to the individual alone but had significance at the 'collective level of a group' too (p.93).

'Spatial practice with all its contradictions in everyday life, space perceived in the commonsensical mode – or better still, ignored one minute and over-fetishized the next'; 'Representation of space (which might equally be thought of as discourses on space); the discursive regimes of analysis, spatial and planning professions and expert knowledges that conceive of space'; and 'Spaces of representation (which might best be thought of as the discourse of space)', the third term or 'other' in Lefebvre's three-part dialectic. This is space as it might be, fully lived space, which bursts forth as what I have called 'moments of presence'. It is derived from both historical sediments within the everyday environment and from utopian elements that shock one into a new conception of the spatialisation of social life' (Shields 1999, pp.160–161). Lefebvre relates the conceptual depictions of space to relations of production. He asks, '*How could the Church survive without churches?*' and explains that social practice relates individuals to a particular space, with a certain level of spatial 'competence' and 'spatial performance' by individuals (Lefebvre 1974, p.55, cited in Shields 1999, p.162).

Social space 'has a part to play among the forces of production' and is 'politically instrumental' as it facilitates 'the control of society while at the same time being a means of production by virtue of the way it is developed (already towns and metropolitan areas are no longer just works and products but also means of production, supplying housing maintaining the labour force, etc.)', among other things (Lefebvre 1991, p. 349). Space itself makes visible the relations in the capitalist state. In its 'ascendant phase', the 'belle epoque', the early distinction 'where bourgeois live on the lower floors, and workers and servants in the garrets', gives way to 'peripheral neighbourhoods' (ibid, p.316). Lefebvre traces the origin of the concept of housing to this period and in effect links it to the spatial differentiation and separation from the rest of the city, and quantified in modular units, and what was defined, he states, 'was the lowest possible threshold of tolerability'. As in the twentieth century, the slums disappeared, and the suburban detached houses contrasted with 'housing estates' just as sharply as the earlier opulent apartments with the garrets of the poor above them (ibid). Here, again, the '*bare minimum*' was in view. Lefebvre's critique of these new working class homes is not only that they form the 'lowest possible threshold of sociability' – the point beyond which survival would be impossible – but also that at this point all social life would have also disappeared, and boundaries will have been defined, separating people in zones – those of 'lowest common denominator' versus those where people 'could spread out in comfort and enjoy those essential luxuries, time and space, to the full' (ibid, pp.316–317). These boundaries are '*fracture lines*' that define the contours of '*real*' social space '*lying beneath the homogeneous surface*'. Fragmentation and separation are the means for the covering up of the 'moral and political order'. Here 'state-bureaucratic order,… simultaneously achieves self-actualization and self-concealment' (Lefebvre 1991, p.317). Lefebvre's writing centres on social space and the city. If the way the city is formed, its zoning, separating and fragmenting the inhabitants; affect its self-actualising (at the same time as self-concealing), then the effects of the alienation of the labourer might be seen similarly in the way social relations within the home are played out, or merely the way we function in our homes. Can the individual self-actualize in the context of her home, or does the dynamic of the state's actions and policies make that impossible?

4.7 Self-Agency and Dis-alienation

How we occupy space and how we change it may have the potential to lead to '*dis-alienation*', as an act of self-agency. Where alienation is linked to disempowerment, agency may be linked to autonomy and empowerment and understood as the capacity of the individual to 'make a difference' to a 'pre-existing state of affairs or course of events' (Giddens 1984, cited in Schneider and Till 2009).

In order to find the condition in which a process of self-agency may emerge out of alienation, we may look to dialectics. These opposites relate to each other dialectically. Lefebvre explains his concept of *spatial practice*, on the basis of the construction of a 'spatial code' and the idea that 'an already produced space can be decoded, can be read'. The 'rise, role and demise' of 'a particular spatial/social practice' and its corresponding 'codifications' will have been produced along with the space corresponding to them. This, Lefebvre explains by the dialectical character of their relations (Lefebvre 1991, p.17). Alienation and dialectical relationship between phenomena are important parts of the tools for understanding the spatial production of the city, and it is possible that these can be used as tools to understand the spatial relationships in the interior of dwellings.

The productive unit that was the family and lived and worked in the dwelling had a social dimension; the current dwelling is regaining that partly with the live-work developments and with the increase in working from home which has been made possible by the advances in digital technology. The 'spatial practice of a society' may be related to the idea of agency as a means of transforming (Schneider and Till 2009) and of making and doing (Turner 1972). Turner's idea of 'housing as a verb' deals with the matter in concrete terms and proposes that people should be able to create their own dwellings – a notion that is also evident in the Walter Segal method of construction. However, the limitations of these proposals are obvious, given the limited access to land and to the skills required for construction.

The capitalist mode of production separates and fragments our experiences. Just as our disconnection with nature becomes a part of this fragmentation, appropriating it is our best chance at gaining some control. 'The more a man is able to take possession of the outside world – through his senses, his spirit, his intelligence – and the more integrated and many-sided this taking of possession, this 'appropriation' is, the greater is his chance of becoming a whole man' (Fischer 1975, p.23).

To be whole we need to enact our agency (and autonomy) and to interact with our immediate environment. Commenting on the importance of human interaction, Doreen Massey argues, 'if time unfolds as change then space unfolds as interaction… space is the social dimension… it is the sphere of continuous production through practices of material engagement' (Massey 2005, cited in Loxley et al.

2011, p.47). Our experience of the home has been linked to memory (Bachelard 1964) and of projections (Cieraard 2015). In both of these variations, it appears that the imagined takes precedence over the actual. However, seeing the two in terms of their dialectical relationship, the relationship between the actual and the imagined and vice versa becomes more of an enabling force than a passive one. In this context the memories and projections may be seen as components of the self and self-agency. In her study, Cieraad found that '…those individuals who share memories of home are more likely to create the same projections of future homes' (Cieraad 2015, p93). Rather significantly, her study found that '[T]he intertwinement of collective memories and projections is most evident in the case of migrants and refugees'. It is interesting to note the inclusion of the latter group, who are in a most contradictory position possible in terms of self-agency. Cieraad's study of the condition of homeless people is interesting in this sense too. The study found that homeless people 'tend to have the most romantic fantasies about a future home with loved ones, as if to compensate for the misery they are in' (Moore 1994, cited in Cieraad 2015, p.93).

We find meaning through our social roles, and our position within the modes of production determines this to a large extent. Our alienation is not limited to the fragmentation of our agency from the product of our labour but also the fragmentation of our connections with our immediate social networks and the larger networks. The design of homes can alter the level of our sense of agency just as technological innovations alter our means of relating to others and to the environment. To lift housing to something more humane that at 'the threshold of tolerability', it is relevant to know how we can design for greater self-agency for the occupants, to discover how we might recognize and measure self-agency and how can we mitigate the effects of alienation. It is proposed here that the answer must be sought in our enhanced connectivity and self-agency and tested through empirical research that seeks to observe, analyse and understand the concept of self-agency as it is played out in the home as well as in the urban space.

Bibliography

Bachelard, G. (1964). *The poetics of space*. Boston: Mass.
Bourdieu, P. (1977). *Outline of a theory of practice*. Cambridge: Cambridge University Press.
Cieraad, I. (2015). Homes from home: Memories and projections. *Home Cultures, 7*(1), 85–102.
Cuboniks, L. *Xenofeminism: A politics of alienation*. Available from: laboricuboniks.net. Accessed 27 Apr 2016
De Certeau, M. (1984). *The practice of everyday life*. Berkeley: University of California Press.
Dictionary (2017). Available from: https://en.oxforddictionaries.com/definition/alienation. Accessed 16 Aug 2017
Edwards, M. (2016). The housing crisis and London. *City, 20*(2), 222–237., Routledge Taylor Francis.
Fischer, E. (1975). *Marx in his own words*. London: Pelican books.
Foucault, M. (1972). *Power/knowledge*. New York: Pantheon Books.

Giddens, A. (1984). *The constitution of society: Outline of the theory of structuration*. Berkeley: University of California Press.

Gorz, A. (1989). *Critique of economic reason*. London: Verso.

Habermas, J. (1983). Modernity: An incomplete project. In H. Foster (Ed.), *Postmodern culture*. London: Pluto Press.

Harvey, D. (1982). *The limits of capital*. Oxford: Oxford University Press.

Harvey, D. (1990). *The condition of postmodernity*. Cambridge: Blackwell.

Harvey, D. (2002). *Spaces of hope*. Edinburgh: Edinburgh University Press.

Harvey, D. (2010). *The Enigma of capital and the crises of capitalism*. London: Profile Books.

Harvey, D. (2015). *Seventeen contradictions and the end of capitalism*. London: Profile Books.

Lefebvre, H. (1991). *The production of space* (trans. Nicholson Smith, D). USA, UK, Australia: Blackwell Publishing

Lefebvre, H. (1974). *La Production de l'space'*. *L'Homme et la societe* 31–32 (January–June), pp. 15–32

Loukola, O., Kyllönen, S. (2005). The philosophies of sustainability in: Anneli Jalkanen & Pekka Nygren (Eds.), Sustainable use of renewable natural resources — from principles to practices. University of Helsinki Department of Forest Ecology Publications 34. Available from: http://www.helsinki.fi/metsatieteet/tutkimus/sunare/21_Loukola_Kyllonen.pdf. Accessed 17 Aug 2017

Loxley, A., O'Leary, B., Minton, S. (2011). *Space makers or space cadets? Exploring children's perceptions of space and place in the context of a Dublin primary school*. Educational and Child Psychology, 28(1), 46–63.

Marx, K. (1844). *Second manuscript- human requirements and Division of Labor Under the Rule of Private Property and Under Socialism* (p. 30). Division of Labor in Bourgeois Society. Available from: https://www.marxists.org/archive/marx/works/download/pdf/Economic-Philosophic-Manuscripts-1844.pdf. Accessed 24 Jan 2016

Marx, K. (1975). '1844 Economic and Philosophical Manuscripts'. *Early Writings* (pp. 322–30) London: Penguin.

Massey, D. (2005). *For Space*. London: Sage, cited in Loxley, O'Leary, Minton.

Moore, J. (1994). *Home: Image or Reality? The meaning of home to 'Homeless People'*, Paper presented at Ideal Homes? Conference, Teeside University

Morrison, K. (1998). *Marx, Durkheim, Weber, formations of modern social thought*. London: Sage Publications.

Office for National Statistics (2014). *Characteristics of home workers*, Home working dcp171776_365592.pdf. Accessed 21 Aug 2016, www.ons.gov.uk

Office for National Statistics (2015). Table FT1101 (S101): Trends in tenure. Accessed 21 Aug 2016, www.ons.gov.uk

Schneider, T., Till, J. (2009, Spring). *Beyond discourse: Notes on spatial agency*, Footprint, Agency in Architecture: Reframing Criticality in Theory and Practice

Shields, R. (1999). *Lefebvre, Love and struggle, spatial dialectics*. London/New York: Routledge.

Turner, J. (1972). *Freedom to build: Dweller control of the housing process*. New York: Macmillan.

UN (2016). Universal Declaration of Human Rights, Article 25, Available from: http://www.un.org/en/universaldeclaration-human-rights. Accessed 28 Nov 2016

UN (2017). The Sustainable Development Goals Report 2017, Foreword by Antonio Guterres, Available from: https://unstats.un.org/sdgs/files/report/2017/TheSustainableDevelopment GoalsReport2017.pdf. Accessed 16 Aug 2017.

Wendling, A. E. (2009). *Karl Marx on technology and alienation*. UK: Palgarve, Macmillan

Chapter 5
The Misalignment of Policy and Practice in Sustainable Urban Design

Michael Crilly and Mark Lemon

5.1 Introduction

A so-called urban renaissance throughout the UK's major cities began in the late 1990s as a policy response to the dominant trends in counter-urbanisation and inner-city decline. This urban policy encompassed the interconnected themes of sustainability and design quality as the basis for public sector investments into housing and urban infrastructure.

This chapter explores these ideas through a chronology of urban regeneration and neighbourhood planning policies and, in so doing, tracks the progression of sustainability and design quality and community engagement in the planning system and their involvement in urban redevelopment and supporting policy. These themes are presented from the perspective of an urban design practitioner who has spent much of the past 20 years working through the changing policy landscape. In consequence, it is intended to provide a different sort of a review of sustainable urban regeneration in England, one that is based on a practical knowledge of urban planning policy and personal experiences of the different national regeneration programmes, particularly as they have impacted on the North of England.

This chapter will present the lead authors' case study materials to highlight the implementation gap between the intent of urban design policy and the actual impact on the ground. It charts observations of unforeseen consequences, the changing targets and policy definitions, including the policy trajectory of sustainability and quality within the built environment and the move from centralisation towards bypassing local government to the promotion of community-led development.

M. Crilly (✉)
Studio Urban Area LLP, Newcastle upon Tyne, UK
e-mail: michael@urbanarea.co.uk

M. Lemon
Institute of Energy and Sustainable Development, De Montfort University, Leicester, UK

© Springer International Publishing AG, part of Springer Nature 2018
M. Dastbaz et al. (eds.), *Smart Futures, Challenges of Urbanisation, and Social Sustainability*, https://doi.org/10.1007/978-3-319-74549-7_5

5.2   The Challenge for Cities

The genesis of the so-called urban renaissance in the UK was planted in the Garden Festivals and broader central government City Challenge programmes of the late 1980s and early 1990s. Set against a decline in traditional industries, and a significant and measurable population movement away from the inner city and towards suburbia, the governments of Margaret Thatcher and John Major initiated a number of urban policy programmes to arrest the decline in population occurring in most of the northern conurbations. This was predominantly through the drive of economic restructuring towards a service- or knowledge-based economy that, in practice, equated to a variety of out-of-town retail parks and a preponderance of customer call centres (Charles and Benneworth 2001).

Indeed, at the time, stopping the dominant trend of counter-urbanisation seemed almost unachievable as it was underpinned by economic decline and a complex raft of associated social problems. There was a debate around the ability, if not the commitment of successive governments, to effectively challenge and change these social trends. In an extract from a confidential letter to the Prime Minister Margaret Thatcher which was released from the National Archives (2011), Chancellor of the Exchequer Geoffrey Howe, in referring to the city of Liverpool, wrote in August 1981 that 'I cannot help feeling that the option of managed decline … is one which we should not forget altogether. We must not expend all our resources in trying to make water flow uphill'. Mr. Howe emphasised the sensitivity of the issue a month later, when he wrote that "'managed decline' ... is not a term for use, even privately". As the Chancellor, he was basically questioning whether going against the trends was not just a waste of money.

Given the contrasting political allegiances of national and city governance, with most of the metropolitan and urban authorities in the north being Labour controlled but working in the context of the Conservative administrations, there were consistent suggestions of an implicit strategy for continued managed decline rather than investment and urban regeneration. Unfortunately in the early 1990s, centralised strategies for urban design and regeneration were already suffering from a healthy level of political scepticism, particularly from local city councillors' view of some of the earlier regeneration programmes (Beecham 1992). Seemingly the baggage attached to many programmes, because of the party political associations with those who initiated them, meant that they maintained an odour of unacceptability and a hint of bias towards a suburban rather than an inner-city and urban electorate.

At the time the *process* of urban regeneration was largely secondary to the anticipated *substantive* outcomes of any investment, as were any considerations around sustainability. Yet much of the policy and academic response to the actions of the *City Challenge* programme (Robinson 1997) and the more joined-up follow-up *Single Regeneration Budget* (SRB) programme stressed that sustainability is about process (Smith et al. 1999) and community involvement, capacity and empowerment (Chanan et al. 1999) rather than a technical issue. Yet, there was also an apparent consensus on the key challenges of stopping counter-urbanisation and addressing the poor image and stigma attached to inner-city areas and the existing social housing stock across many of the northern conurbations. 'Most cities were losing population … (i)n some cases, there was simply too much housing in the wrong places. Inner city estates were in some ways easier to restore than outer estates. They were closer to shops, transport, jobs and other services. But breaking up these estates and blending them into the urban surroundings was always expensive and difficult. They often continue to stand out as "council housing" even after exceptional spending to integrate them' (Power and Tunstall 1995 p71–72). It was within this context and the assessment of the urban policy efforts to date that the procedural challenges of urban regeneration were carried forward into the membership of a newly formed Urban Task Force.

5.3   Towards an Urban Renaissance

There was a great excitement in local government planning when the new Labour government initiated their Urban Renaissance, particularly with the appointment of Lord Rodgers of Riverside to lead a task force and compile a report and recommendations. At the time, Rodgers was not so well known outside of the world of architecture, compared to his wife Ruth who had established and ran the River Café underneath his architectural office complex on the north bank of the Thames. Indeed, the printed Urban Task Force report borrowed much inspiration from the River Café Cookbook, in its shape, layout and contrasting mustard and plum colours and in its contents. It provided a set of ingredients and recipes for cooking up urban regeneration.

There was a significant pro-urban theory underpinning this new Urban Task Force, based on a polycentric city, understanding of mixed use communities, revised assumptions about how increasing urban density and integration with public transport has the potential to deliver a more sustainable urban form. This was a newly discovered conceptual model to guide policy intervention. It demonstrated a complex analytical model of understanding the entire conurbation/city, region and communities within the wider urban system through process models and substantive/physical examples of sustainable communities, perhaps seeding sustainable urban design as an integration between physical and social infrastructure and networks. In short, it started to explain what an integrated response to the urban challenge might look like.

However, what also appeared was a top-down critique of current practice and a plea for investments and interventions to have a strong evidence base. At the national level, this urban programme was supported by the creation of the Commission for Architecture and the Built Environment, the Commission on Sustainable Development and the Commission for Integrated Transport, amongst others, as suitable quangos to deliver the necessary training, support and, where necessary, direct strategic intervention into the delivery of the programme. This was often repeated at a regional and local level with investment in design quality and sustainability through a network of Architecture and Built Environment Centres of Excellence. These mixed organisations' self-accepted responsibility for building capacity and a skills base within local authorities and communities and the desire for an underlying evidence base for policy and capacity-building was to become a repeating theme over the next decade.

Yet, in a strangely contradictory way, individual members of the Task Force had already used their evidence base to make political and locality specific criticisms of current regeneration and planning practice in both Manchester and Newcastle upon Tyne. This was evident in the number of thinly veiled references to recent greenbelt releases and the anonymous references to the declining inner-city areas of Elswick, South Benwell, and Scotswood (aka Bankside and Riverview in Power and Mumford 1999 p48–62). 'Bankside has huge potential. Its housing is not only adequate, it is excellent'. ... 'The city has plans to extend onto the greenbelt because people do like to move out. But we have a site here that if it were ten miles further along the river would be worth millions. It's south facing, it's sunny and the views are stunning. There's no capital being made of the location'.

In this review, the challenge of sustainable urban regeneration was being explicitly linked to greenfield releases and suggestions that '(b)oth Manchester and Newcastle have been hard hit by sprawling greenfield building. ... The statement within Newcastle's Plan that the city still suffers from housing shortages and too high density contradicts available evidence.' (Power and Mumford 1999 p73). In a partial rebuttal to this critique, the leader of Newcastle City Council at the time made it clear that urban regeneration is difficult if it 'ignores the market' (Flynn 2000), and the challenge was to deal with the area-wide stigma created, not so much by the physical environment as the levels of crime and poor services (Blackman 2001). This required a package of measures and interventions from a variety of stakeholders (Cole et al. 1999) in as coordinated a manner as possible.

These academic references in turn became part of the critique of the Urban Task Force itself, which did '... not accept the argument of certain northern planning authorities that the way to overcome low demand for housing in their area is to build on the surrounding greenfields, rather than tackling the regeneration of their urban heartlands. The release of such land will simply exacerbate their long term problems' (Urban Task Force 1999 p217–218). In making these arguments against piecemeal decision-making, particularly the separation between housing and other aspects of social investment, health, education and crime, they were highlighting how difficult inner urban regeneration would be, particularly when being hit by excessive amounts of suburban greenfield housing development at the same time.

Central to this challenge for sustainable regeneration was the need to address the perceived market and find a way to shape or change attitudes to urban living, to start marketing and branding the attractions of urban living in a more imaginative and creative way. One attempt to do this while informing the Urban Task Force was an influential and important report by MORI (et al. 1999). This was important in part because it coined the phrase 'urban pioneers' as the first potential wave of regeneration. It raised expectations that creative households would be the first to be attracted to a sustainable and compact urban form that provided more bespoke choices for living in comparison to new housing on the outskirts of the city. It extended earlier work that demonstrated how views of cities and city living could change for the better, particularly how Manchester (Hebbert 2010) led the rise of city centre living and set significant precedent for other northern conurbations seeking to add additional choice and unique property typologies to the housing market (Chesterton 1998). This was supported by similar findings in qualitative research looking at different attitudinal groups such as 'committed residents', 'budding incomers' and 'probable leavers' within social housing areas (Dean and Hastings 2000) and more localised challenges within individual urban housing markets (Crilly et al. 2004, Townshend 2006), all in an attempt to understand the interrelated and complex issues of the housing property market within urban regeneration.

So in an appropriately suitable quirk of fate, and a response to the need for strong corporate and civic leadership, Richard Rodgers's firm was invited and appointed to work with Newcastle City Council to bring this thinking to Tyneside and jointly produce the 'Going for Growth' regeneration plans for the west end of the city (Newcastle City Council 2001). This district scale masterplan set out the expected polycentric model of revitalised urban communities throughout the west end of the city, supported by investments in green infrastructure, retail and community facilities. The plan also included a feasibility testing of an extension to the Tyne and Wear Metro system into the area as a network of street running trams. In addition, it set out options for significant, albeit targeted demolition and the creation of a new replacement urban village at Scotswood, an area on the river about two and half miles west of the city centre.

However, the response to this first wave of urban regeneration work in Newcastle, post Urban Renaissance and Task Force recommendations attracted familiar and severe criticism of the *process* (Walker 2002; Newcastle Unison 2002) as much as the *substantive* proposals, with the key critics highlighting the 'selective involvement' with existing community groups. From the outset, reviewers suggested that the well-intentioned corporate strategy would have many negative unexpected consequences relating to forced gentrification and the lack of coordination between housing and the critical element of schools. The regeneration agenda was confused at the local level through a mix of misdirection, misinformation and resultant misunderstandings. Responses were driven by mixed emotions, including some deliberate and politically motivated sabotage. Many of the reviews of this renaissance programme were politically influenced (Shaw 2000) and factually incorrect as a result, often, and ironically, missing the significance of the underlying evidence base used to inform decision-making.

This initial 'Going for Growth' regeneration strategy in Newcastle had press and researchers charting the adverse reaction to the radical strategy and the disaffection of the top-down process (Cameron 2003; Byrne 2000) with the view that '... the council has done the dirty on all the residents' associations' (Young and Dickinson 2000 p8). However, similar to much sociological research based on this regeneration strategy, there was explicit criticism without any clear changes or solutions being advocated beyond the basic platitudes of closer partnership working and community ownership of the proposals. These reviews (Coaffee 2004) ignored the more fundamental issues of the necessary resources to make this work in practice. One interesting characteristic of this strategy was the imbalance in resources around delivery against external research—at one point there was up to six universities and twenty different research teams working on aspects of the schemes—far in excess of the resources given to supporting the delivery of the programme itself. Herein there was a practical paradox between resourcing community engagement and empowerment with promotion of the prevailing national pro-urbanist compact sustainable city agenda. Retrospectively it was clearly useful to understand Urban Task Force-guided local actions within the longer-term perspective of urban regeneration policies (Wilks-Heeg 2000), not least because of the continuing involvement of the same organisations and individuals. In part, this reaction to what was perceived as a demolition and gentrification strategy was more about maintaining the strong community networks (BKW Tenants & Residents Association 2000) that existed in the proposal areas than it was about the actual value of the bricks and mortar. The rationale behind many objections was to prevent the community being destroyed alongside the houses; it was about trusting the motives and the authors behind the proposals. It was also clear that the scale of addressing regeneration had to be at the city-region scale and thus required a different sort of collaborative coordination between multiple local authorities and agencies. It was never something that Newcastle City Council could have achieved on their own as a corporate body.

At first glance, central government appeared to be making the same, albeit more generic, points about the importance of effective community government (ODPM 2002) in setting out their position of community involvement. It was saying that coordination and integration of community involvement was a prerequisite for sustainable regeneration. It was also saying that the process being followed was as critical as the substance of the programme while suggesting that the role of existing local councillors with electoral mandates was not sufficient to represent the diverse and more localised urban communities.

5.4   Renewing the Housing Market

Putting this urban renaissance more firmly into national policy, John Prescott the Deputy Prime Minister (ODPM 2003a) understood that 'low demand requires a new approach, to recreate places where people want to live—not leave. This means tackling not just housing but where we can, rebuilding sustainable communities'.

Thus, when resources finally came to support the urban regeneration process, it was largely through the *Housing Market Renewal* (HMR) *Pathfinder* programme. This was part of the wider Sustainable Communities Plan (ODPM 2003b), a much larger and more comprehensive programme of investment into southern growth areas and decent 'social' homes which explicitly linked investment to sustainability and design quality, particularly at the scale of the neighbourhood. The *HMR Pathfinder* programme was arguably the best- resourced urban regeneration programme of the last 15 years and was based on significant levels of housing demolition and clearance in areas of low demand. It was intended to address the poor quality of much of the existing older housing stock together with a rebalancing of the housing choices available within local markets in the northern conurbations.

Yet almost from the outset of the programme, there were concerns over the lack of clarity and coherence around the strategic objectives. There were warnings coming out of different central government departments, questioning the focus of the programme and the balance between demolition and refurbishment. Much of the 'quick fix' in the plans was about the demolition of unpopular housing, as a preparatory stage in the larger programme, rather than understanding the underlying complexity of urban attitudes and the housing preferences needed for the delivery of sustainable and affordable housing.

Also underlying the advocacy for better alignment and less single issue programmes (House of Commons 2003), in part a political nod towards maintaining the *Single Regeneration Budget*, there was pressure to include a real statement regarding resources, with regeneration activities requiring continuity over ensuring long-term funding for long-term projects, avoiding the unequal distribution of resources which is often seen as politically divisive and getting resources for better skilled and supported professionals (ODPM 2004a).

In mid-2003, early into the *HMR Pathfinder* programme, the Commission for Architecture and the Built Environment (CABE) made a statement on behalf of all of the national agencies, saying '… (w)e have an opportunity that cannot be squandered; the policies and proposals of this government could impact around one million homes and the communities that live in them … (n)ot since the 19th and early 20th century has England had to deal with housing renewal on the scale envisaged for this programme. We have a once-in-a generation opportunity to get it right' (CABE 2003). This included the promotion of strategic advice on design quality, sustainability and, most significantly in practice, issues around the processes and procedures to follow to ensure quality and sustainability standards (CABE et al. 2003). In this context, there was a close national alignment between design quality and sustainability standards. CABE and others were saying that housing demolition should be a last resort and only undertaken after replacement proposals of higher-quality and sustainability standards have secured statutory planning approval and with funding for delivery already in place.

It was correct in identifying a central role for masterplans, particularly those that were better aligned with statutory plans and longer-term thinking. In a lot of ways, this was simply restating in a more forceful manner, earlier recommendations (CABE et al. 2003) that decisions are based on robust evidence and market analysis (NAO 2007a), supported by clear objectives within approved masterplans. This advice and support certainly supported more advisors and a rapid growth in the masterplanning industry.

There was a delayed reaction to these concerns, with recommendations that top-down government support was to include qualitative considerations into the policy scope for sustainable communities and not just to be about energy use (House of Commons 2005, ODPM 2005). Yet, short-term outcomes were being driven forward at the cost of adequate consultation and engagement processes. There was also too much geographical focus on tightly defined boundaries. These boundaries limited the potential of wider northern conurbations to deal with areas of low demand while failing to address the structural imbalances of housing demand between the midlands, the north and the south-east growth areas.

The delivery of the *HMR Pathfinder* programme alongside the *Growth Areas* then began a rethink of the processes for housing delivery. Indeed, the balance between affordability and sustainability had been central to government policy leading into the 2007 Housing Green Paper (DCLG 2007), and throughout this chronology of policy there have been the repeating themes of quality, sustainability and affordability. So often 'best practice' precedents were sought to demonstrate firstly how these themes could be shown to be achievable together and secondly that the developing planning and development policy could in effect be internally consistent. This was evident in the launch of the *Code for Sustainable Homes* as a new national standard for zero carbon development. This was a standard that was tested at the strategic scale of the sustainable community in high-profile national initiatives such as the *Carbon Challenge* and the *Eco-Towns* initiative, each designed to demonstrate the delivery of zero carbon targets in practice by working in partnership with commercial developers. Yet, the tensions between raising standards and rapidly delivering housing numbers; or failing that, large numbers of demolition; remained unresolved.

In practice, the attraction of additional funding seemed to be too much of a draw for *HMR Pathfinder* bodies, alongside a second wave of authorities. Areas that were excluded from the first round of regeneration funding perceived themselves to be in competition for limited central government resources. At the time, many authorities considered it would be better to secure the funding through rapid action rather than long-term considered regeneration programmes.

For example, the second wave of pathfinder funding going to Teesside was initiated on the basis that 'HMR must not be seen as purely a clearance issue (para 12) … and that detailed collaborative masterplan and costed options to be created but led by the promise of central government funding' (Middlesbrough Council 2005a). The suspicion that the *HMR Pathfinder* was largely a funding-led approach originated when their consultant's report stated that '…(t)here is the prospect of major housing renewal resources for Middlesbrough's Older Housing, if the strategy for tackling it is right' (Nathaniel Lichfield and Partners 2004 p63).

This race for funding was in direct contrast to the policy objectives, where '(t)he objective for all the proposals is to firstly understand and establish the particular issues of the area concerned and endeavour to direct the market in these areas to a more sustainable position, removing their reliance upon public funding' (Chapman 2004 p32). Statutory requirements for local authority use of compulsory purchase powers for the purposes of land acquisition and assembly of development sites included the consideration of all viable options that implicitly included sustainable refurbishment. Yet sustainability, energy efficiency and the consideration of refurbishment as well as demolition and replacement were often ignored within practicable decision-making in direct contradiction to many strategic policy requirements. There was little government interest in addressing the sustainable refurbishment of the existing housing stock through financial incentives (SDC 2005). This leads to several high-profile examples, particularly in Liverpool and the Tees Valley, of areas being targeted for demolition without looking at any alternatives.

Nationally, there were deep political concerns expressed (House of Commons 2005) that the programme reflects a demolition (Hetherington 2005) rather than a regeneration strategy, with '… criticisms (that) appear to confirm the worst fears of conservation groups, who claim there is no justification for pulling down historic terraces (and replacing) with poor-quality new homes' (Weaver 2005).

So, belatedly in the programme, design quality and sustainability standards were more forcefully introduced as a prerequisite for financial support for new development, including most of the high-profile exemplar projects. These standards were based on the accepted advocacy of CABE and others. In effect this was a policy acknowledgement of the findings of the underlying evidence base that drew on the prevailing set of collective standards being used throughout the planning system, albeit that the collective set of standards being adopted by the *HMR Pathfinders* was now recognisably that being used by central government agencies and as part of the 'common language' being promoted throughout the design and construction industry (NAO 2007b, Callcutt 2007). Throughout this period, the top-down imposition of quality and sustainability standards prevailed. These standards were defined at the national level, were restricted on the basis of practicality to quantitative measures that relate to individual houses and were imposed as a condition of public-sector funding rather than through local statutory planning. So outside of any funding agreement, they were seen and understood simply as advisory standards and thus largely ignored by the development industry.

Yet these quality and sustainability standards did not always change the actions undertaken as part of the *HMR Pathfinder* programme. Within the Hull regeneration area, the major problems being faced were described as structural within the housing market. Specifically, policy research suggested that there was a lack of diversity and resilience in the mix of types and tenures (Gateway Pathfinder 2006). The regeneration area was characterised by large areas of homogenous housing, with regard to age, size, condition and tenure. A continuation of these characteristics and trends was assumed to make the area housing market vulnerable, and thus demolition continued on the basis of a homogenous local housing market. Similarly, in Teesside, underlying concerns appeared to have been generally accepted in that there was an imbalance in supply and demand that was having a negative impact on the town's housing market. Yet, in reality this justifying evidence was crude and often 'anecdotal' (Middlesbrough Council 2005b para 18) with respect to local geography and spatial distinctions and guess work around the costs of stock improvement (para 25). Indeed, in spite of bespoke research highlighting that '(t) here is insufficient comprehensive research to the carbon emissions of refurbished housing compared to the emissions from new housing when considering the requirement to demolish the existing stock' (CABE Tees Valley 2007 p1), several authorities within Teesside utilised arguments around sustainability and reduced carbon emissions from 'potentially' better quality replacement new housing to promote large-scale demolition of the older housing stock. Even when these arguments lacked supporting evidence, they still progressed to several high-profile public inquiries and became the basis for compulsory purchase orders needed for large-scale demolition.

And similar to the early criticisms of the regeneration work in Newcastle, the predominant fear in Middlesbrough was also about the destruction of existing communities, with high-profile professionals employed to undertake the work (Barrie 2009) stressing the value in strengthening and maintaining sustainable social networks as a precursor to effective urban regeneration.

The justification of low demand leading to proposals for demolition meant that it was being measured against property value or prices rather than affordability. This drew on planning methods from higher-level spatial planning, such as regional spatial strategies and strategic housing market analysis. Thus, in practice properties were being targeted in many cases because they were simply 'affordable'. Areas of homogenous housing were being targeted because they did not demonstrate the diversity and mix of types and tenures that were being advocated in these strategic planning documents. In many instances, people have had their properties compulsory purchase and have been unable to use the purchase cost to find any similar property that is affordable. It had the impact of increasing household bills (Ambrose and MacDonald 2001) and forced some families into poverty (McHardy 2001) as rents increase and other unexpected charges appear.

Indeed, the associated timescale for project delivery meant that planners were unable to address the existence and extent of supporting evidence used to both justify and inform decisions. The evidence-based approach to regeneration investment failed, due to the time delay between stated policy objectives, commissioned evidence base and the creation of supporting research. Much of the required rushed research and masterplanning was a simple regurgitation of old knowledge in new clothes. Quantitative materials dominated at the cost of understanding the importance of design quality, sustainability and the potential to influence household attitudes on urban living.

Towards the end of the programme, many communities that had experienced multiple well-meaning waves of regeneration, building up to the *HMR Pathfinder* tsunami, suffered from 'regeneration fatigue' (Armstrong and Pattison 2012). It seemed that the disconnection between different parts of the *HMR Pathfinder* programme and different delivery teams, at all levels of government, meant that there was a gap between the programme ambitions and the anticipated and actual outcomes.

5.5   Finding the Northern Way

In a speech to the Core Cities summit, John Prescott (2003) raised the importance of cities as regional economic drivers, saying that '(y)ou can't have a sustainable community without a strong local economy and the jobs that come with it. Previous government's forgot that. They built houses but they didn't build communities. ... (t) hey forgot about what makes people want to live in cities. They forgot about people'. Prescott thought that '(b)y getting architects, town planners and developers to work together … (it had) … created a new "wow" factor in the North' (ODPM 2004b p5). Yet, this so-called wow factor was underpinned by quantitative evidence around infrastructure and inward investment as much as any evidence about better quality design or sustainability standards. And to continue with this collaborative 'wow' factor, procedurally, regional development agencies (RDAs) and local partnerships remained in the driving seat with local authorities having a limited role alongside community sector, business and central government interests.

Following this, the 'Northern Way' as part of the wider *Sustainable Communities Plan* also embodied the idea of extending devolved power from central government to regional assemblies and regionally elected mayors with more power to coordinate activities. There were also significant roles for new *Urban Regeneration Companies* aimed at focussed and strategic interventions. Again, the proposed approach was to support coordination, integration and partnership working. Yet, in practice, it seemed that there were many political appointments sitting on the board of these agencies and corporations. Partnership working became a mystical process around internal coordination and agreement between the plethora of new agencies. There seemed to be a certain lack of trust of local politicians in allocating regeneration resources through this pick and mix approach to partnership building.

The financial crisis of 2007–2008 and the subsequent general election of 2010 started to change things. The crisis around high-risk mortgages started at *Northern Rock* and their headquarters in the Newcastle's suburb of Gosforth. In the midst of the first run on a bank in over 150 years, the underlying assumptions about the housing markets in the northern conurbations changed radically. In a period of growing austerity that seemed to affect the north most dramatically with its higher proportion of public sector employees, the 2010 general election removed Labour from power and returned a Conservative and Liberal Democrat coalition government. They wasted little time and little thought in halting many of the centrally funded regeneration programmes and organisations removing them in what became known as the 'bonfire of the quangos' (HM Government 2010).

With a change in national government came the inevitable shift in policy emphasis. Firstly, more focus was placed on housing as a sector. The collation government became interested in the delivery of more residential units at a national level. This took place in a context where there had been significant shortfalls in the number of actual housing completions compared with the aspirational targets set against the demand evident in the strategic housing market assessments underpinning statutory planning documents around the country.

They also seemed to be suggesting that local government and the planning system were failing alongside other limiting factors imposed on the housing development sector, including the skills deficit within the construction sector (CITB 2003).

The coalition government's assessment of the original *HMR Pathfinder* scheme as articulated by the Housing Minister Grant Shapps at the time was that '… local communities in some of the most deprived areas of the country were told they would see a transformation of their areas. But in reality, this amounted to bulldozing buildings and knocking down neighbourhoods, pitting neighbour against neighbour, demolishing our Victorian heritage and leaving families trapped in abandoned streets. This programme was a failure and an abject lesson to policy makers' (DCLG 2011). Yet, damming criticism of the programme was not restricted to politicians. Writing about the Liverpool experience, comedian Alexei Sayle said that "… John Prescott's cretinous *Pathfinder* scheme (working with) … the deposed leader of the city council, admitted that their efforts connected with Pathfinder … to compulsorily purchase and demolish thousands of 19th century homes had left many communities 'looking like warzones'" (2012 p30). Localised anger around the plethora of poor outcomes from the *HMR Pathfinder* programme was almost universal, and what each individual programme team perceived in isolation to be NIMBY-based resistance to change was, in actuality, the dominant community response to the ill-thought-out programme throughout the country. Alongside multiple admissions, or accusations, of failure at national and local levels was a fresh funding initiative to clear up the mess.

The resultant planning and regeneration blight is now being dealt with in a somewhat reactionary manner by throwing more resources at the perceived problems alongside the (re)setting of familiar metrics for assessing the benefit of spending this additional money. These included considerations such as the number of homes demolished and more positively the number of empty properties brought back into beneficial use. This was driven by an imposed requirement for the Department for Communities and Local Government to consider refurbishment options. Yet again, underlying this allocation of funding is a centralised approach, set out in bidding criteria (DCLG and HCA 2011) for local authorities aiming to produce 'exit strategies' that are targeted at streets or urban blocks that are more than 50% vacant. On investigation, this metric was not derived from any sound evidence. Perhaps more worryingly was the potential for local authorities to 'fix' this figure by defining street and block boundaries to include the footprint of already demolished properties as part of the 50% of properties that can be considered vacant. In this approach to the calculations of the metric, the central government agencies and local authorities were able to manipulate and direct their funding as they felt was required to finish the demolition job.

There were certainly aspects of self-interest for some local authorities in this clean-up funding. It seemed to be the case that completing the job of housing clearances brought benefits in efficiency to the local authority and service partners with no more need for temporary site hoardings, security patrols, refuse collections and similar. It seems so obvious to policy makers that short-term limited capital expenditure to remove a problem is a better value to the public purse than the uncertainty of ongoing revenue expenses. There were also a range of quantifiable benefits and revenue savings arising out of the reduced need, or improved efficiency, in the provision of services (e.g. in the key performance criteria service savings; p5 DCLG and HCA 2011) that were being required in this new period of austerity.

So following a critique of a centralised programme that measured its success in part on the number of housing demolitions was another centralised programme with a similar trend in demolition metrics. This approach seems to be repeating so much recent history for the 'demolition-mad paternalism' of the 1980s housing regeneration that has been described as '… a classic example of an anonymous bungling bureaucracy destroying a living community' (Wates and Knevitt 1987 p51). So more criticism (Waite 2012) emerged, this time of the replacement 'clean-up' programme, yet again around additional demolition proposals in Liverpool, Teesside (Polley 2010) and other locations.

Reflecting on the outcomes which have fallen significantly short of addressing the policy aspirations, it now seems that any potential legacy of the *HMR Pathfinder* programme is the implicit approach to national retrofitting policy, albeit starting at the individual household scale. The basis for a significant level of demolition was on the basis of low demand and the poor quality of the existing older housing stock. Yet sustainability, energy efficiency and the consideration of carbon emissions were largely ignored within practicable decision-making in direct contradiction to many strategic policy requirements.

5.6 Creating the Big Society

Throughout this transition period with the new coalition government, there was also a distinct, and suspiciously ideological, change in the policy emphasis from physical to cultural and social regeneration activities with the corresponding focus on localisation and the growth of codesign as an integral part of the planning and redevelopment process. Although commentators (O'Brian and Matthews, 2015) recognised this shift in political emphasis, it was accepted as one of several sudden and reactionary changes to urban policy. The idea of 'new localism' became a fashionable 'buzzword' as well as an approach to pass decision-making responsibilities from central government agencies to local communities and with this any perceived blame for failures to deliver successful regeneration projects

This is perceived to be as much a move by the government to bypass inappropriate and ineffective local authorities as it was about empowering local communities. It did however maintain an emphasis on sustainable design inasmuch as evidence does show there is a significant overlap, if not direct correlation, between community-led enterprises and the emphasis on sustainability (Seyfang 2010). Yet it does seem to be difficult to then take these local sustainable practices and procedural innovations and niche markets and support them at a larger scale. Devolved regeneration and support for sustainable communities also suggested multiple approaches, options and solutions. Expectations were that a diversity of planning approaches might emerge within the systems and that this 'new localism' would mean less of the top-down imposition of central government policies.

One key shift was the National Planning Policy Framework (CLG 2012) that, while placing sustainable development as a key function of the statutory planning system, also introduced wider requirements that collectively aimed to increase the delivery of housing numbers. This included assessment for '… people wishing to build their own homes' (CLG 2012 para 159 p39) that bypassed local government and placed responsibility and devolved budgets into the hands of local community organisations. The default approach was to approve and support the delivery of what in effect was a watered-down definition of sustainable development.

There was also a reinterpretation of 'best value' benefits for local authority working with the community sector. Devolved services and contracts could underpin co-operative groups and have multiple regeneration benefits (Boyle 2012), albeit it that community sector provision was possibly more attractive because it was cheaper.

The idea of design quality remained on the broad agenda and was supported by a national review of quality and sustainability standards through a review led by architect Terry Farrell. The consensus that assessing and measuring quality could remain objective enough for the statutory planning system to implement some minimum standards remained. However, the standards started to relate to 'evidence-based design' (RIBA 2015) where lessons were being learnt through the life of practical projects, from technical assessment through to post-occupation evaluation.

The devolution of funding around the sustainable retrofitting of individual homes also became part the Big Society focus. Schemes such as the Renewable Heat Incentive, the Green Deal and Energy Company Obligation (ECO) emerged as a means to achieve legally binding carbon-saving (Climate Change Act 2008) and fuel poverty targets (Warm Homes and Energy Conservation Act 2000) which were focused on the older building stock that was responsible for around one third of all carbon emissions. Yet, with poor financial incentives, overly bureaucratic procedures and centralised management and validation procedures, all of these renewable energy and sustainable retrofitting programmes died a death, alongside the idea of the Big Society.

5.7   Investing in the Northern Powerhouse

At the point of writing, the *Northern Way* is being rebranded and turned into a strategic statement of intent and investment strategy that links employment, training, connectivity and some limited investment on cultural initiatives. This *Northern Powerhouse* (HM Government 2016) is presented as a start for an ongoing conversation with stakeholders. Yet the scope of the programme has been reduced to that of strategic infrastructure provision and regional economic development, hinting towards a strong economic bias in any ongoing investment programme.

At the same time, and under pressure from the development industry, central government has prevented local authorities from enforcing anything other than nationally recognised standards of sustainability through the statutory planning system. The justification for this was superficially about simplifying the regulatory environment to support the delivery of housing. Yet under additional commercial pressure, they have simultaneously removed the *Code for Sustainable Homes* as a nationally recognised standard of sustainable design. This was justified on the basis that statutory building regulations are of a sufficiently high standard to make any other standard redundant. In so doing, the implicit definition of sustainable design has been limited to energy in use. Many of the broader social and environmental aspects of sustainable design, and considerations at different scales of intervention, are simply being abandoned.

5.8   Summary

The changing intentions behind urban design policy over the last 20 years have shifted from simple economic benefits to a wider integrated approach to environmental and social benefits and then back to an economic focus. Yet, the challenges of building sustainable communities include the need to maintain a strategic overview of, and broad thematic scope for, planning. This is within an institutional and policy context where the dominant trend is for increasing specialisms and substantive outcomes over process.

In the developing world of urban design policy, there has been a repeating paradox between the requirements for speed and responsiveness in decision-making and the bias towards lengthy statutory processes that maintains consultation periods and opportunities for debate. This mismatch between the timescales required for policy formation and politicised decision-making remains. There is an organisational investment in the time and resources needed to get any meaningful masterplan adopted. This has too often resulted in a misplaced momentum being generated behind the wrong sort of strategy.

The prevalent political confidence, that urban forms of development are inherently more sustainable and will become a marketing attraction for the northern housing market, remains to be proven. Urban planners, architects and designers seem to be able to make people think positively about cities and give it a 'wow factor' but not necessarily to the point of getting people to actually move there. In a similar way, the assumption that sustainability, design quality and innovation, even supported by the right type of marketing, can address locational stigma and compensate for higher density living in neighbourhoods with social problems and comparatively poor educational attainment levels may be slightly naive.

The idea of 'patient money' and thinking about quality and sustainability in the long term has been sidelined in policy. In spite of the extensive and mixed evidence regarding the value of sustainability and good design in theory, this has been unconvincing in practice. The separation of capital and revenue costs arising from following a traditional speculative development model has limited incentives for investment in quality and sustainability for those responsible for initiating development and little influence for those responsible for ongoing management and running costs.

Many of these practitioner concerns have been and are still the basis for discussions around the reasons for why big plans fail. Whatever happens next in the progression of urban policy, there is a historical and procedural context within which any professional or politician works. This context, positive or negative, will have an impact on trust, accountability, cooperation and support ultimately what could be understood as successful regeneration.

The need for an underlying robust evidence base remains strong and can underpin and justify the development of urban policy, even though there has been an empirical bias in the overtly technical and data-driven evidence that has too often misdirected a politically motivated process. Thus, this evidence base has to be broader in scope and include the anecdotal and qualitative views from communities with a real knowledge of the economic challenges they are facing. This is important to balance the often misleading views stated and repeated in much corporate spin associated with many public bodies and organisations hoping to talk up the positive changes of a region or a city undergoing radical restructuring. The evidence has to help see through the hype.

What seems to be repeating too often is the historical emphasis on 'end-state planning', imposed from dislocated ministers in London, and the failure to understand the nature of transition and adaptation. Perhaps we would be better to try and understand the dynamic and procedural aspects of planning and managing cities, accepting the inevitability of mistakes and failure within this process.

Finally, we would suggest that urban design is first and foremost a composite political act. The use of evidence and implicit bias within it, the distribution and criteria-based control of resources, the dominant ideology of the property market controlling regeneration (Webb 2010), the empowerment or disenfranchisement of communities and local councils, the evaluation of sustainability against viability and the 'badging' of successful projects are all small but significantly political acts that have supported the progression of a group of political cronies (Rodgers 2017) as they move from one reorganised regeneration quango to the next.

References

Ambrose, P., & MacDonald, D. (2001). *For richer, for poorer? Counting the costs of regeneration in Stepney*. University of Brighton: Health and Social Policy Research Centre.

Armstrong, A., & Pattison, B. (2012). *Delivering effective regeneration: Learning from bridging NewcastleGateshead*. Coalville: Building and Social Housing Foundation.

Barrie, D. (2009). Regeneration as social innovation, not a war game. *Journal of Urban Regeneration and Renewal, 3*(1), 77–91.

Beecham, J. (1992). City challenge: A sceptical view. *Policy Studies, 14*(2), 15–22.

BKW Tenants & Residents Association. (2000) *'Save Our Streets – Save Our Community' A report by Kenilworth, Beechgrove and Warrington Road Tenants and Residents*. Unpublished report for West City Community Project, Cruddas Park Shopping Centre: Newcastle upon Tyne.

Blackman, D. (2001) "Newcastle West End". *Urban Environment Today,* 13th September 12–13.

Boyle, D. (2012). What the social value bill means for planning. *Town and Country Planning, 81*(3), 161–162.

Byrne, D. (2000). Newcastle's going for growth: Governance and planning in a Postindustrial metropolis. *Northern Economic Review, 30*, 3–16.

CABE Tees Valley. (2007). *Carbon-footprinting housing regeneration: Scoping study*. Newcastle: Ove Arup & Partners.

Callcutt, J. (2007). *The Callcutt review of housebuilding delivery.* Wetherby: Department of Communities and Local Government.

Cameron, S. (2003). Gentrification, housing redifferentiation and urban regeneration: Going for growth in Newcastle upon Tyne. *Urban Studies, 40*(12), 2367–2382.

Charles, D., & Benneworth, P. (2001). Situating the north east in the European space economy. In J. Tomaney & N. Ward (Eds.), *A region in transition: North East England at the millennium* (pp. 24–51). Aldershot: Ashgate.

Chesterton. (1998). *City Centre living in the north: A study of residential development within the City centres of northern England.* Leeds: Chesterton.

CITB. (2003). *Construction skills foresight report.* King's Lynn: CITB.

Communities and Local Government. (2012). *National Planning Policy Framework.* London: Department for Communities and Local Government.

Commission for Architecture and the Built Environment, English Heritage, Sustainable Development Commission. (2008). *Housing market renewal: Action plan for delivering successful places.* London: Commission for Architecture and the Built Environment.

Commission for Architecture and the Built Environment. (2003) "Housing market renewal: 'One chance to get it right', say five government advisors". Commission for Architecture and the Built Environment *Press Release* on behalf of English Heritage, Commission for Integrated Transport, Sustainable Development Commission, Environment Agency, 19th June.

Commission for Architecture and the Built Environment, English Heritage, Commission for Integrated Transport, Sustainable Development Commission, & Environment Agency. (2003). *Building sustainable communities: Actions for housing market renewal.* London: Commission for Architecture and the Built Environment.

Chanan, G., West, A., Garratt, C., & Humm, J. (1999). *Regeneration and sustainable communities.* London: Community Development Foundation.

Chapman, J. (2004). The housing market renewal process. *Urban Design, 92,* 32–33.

Coaffee, J. (2004). Re-scaling regeneration: Experiences of merging area-based and city-wide partnerships in urban policy. *International Journal of Public Sector Management, 17*(5), 443–461.

Cole, I., Kane, S., & Robinson, D. (1999). *Changing demand, changing neighbourhoods: The response of social landlords.* Sheffield: Centre for Regional Economic and Social.

Crilly, M., Charge, R., Townshend, T., Simpson, N., & Brocklebank, C. (2004). *NE1 want to live here? Shaping attitudes to urban living and housing options in Newcastle Gateshead.* Newcastle upon Tyne: Newcastle Gateshead Housing Market Renewal Pathfinder.

Dean, J., & Hastings, A. (2000). *Challenging images: Housing Estates, stigma and regeneration.* Bristol: The Policy Press and the Joseph Rowntree Foundation.

Department of Communities and Local Government and Homes & Communities Agency. (2011). *HMR transition funding – DCLG/HCA bidding guidance 1st June.* London: Department of Communities and Local Government / Homes and Communities Agency.

Department of Communities and Local Government. (2011) "£71 million to end the legacy of England's Ghost Streets" November 24th. Available online http://www.communities.gov.uk/news/corporate/2036698 (Accessed 3rd Jan 2012).

Department for Communities and Local Government. (2007). *Homes for the future: More affordable, more sustainable. Cm7191.* Norwich: HMSO.

Gateway Pathfinder. (2006) "Quality streets and much more with design guide". *Pathfinder Progress: The magazine for partners in Hull and East Riding's housing Pathfinder,* Summer 3.

Hebbert, M. (2010). Manchester: Making it happen' pp 51–67. In J. Punter (Ed.), *Urban Design and the British Urban Renaissance.* Abingdon Oxon: Routledge.

Hetherington, P. (2005) "MPs warn against plan to demolish 200,000 homes". *The Guardian,* April 5th p. 6.

HM Government. (2016). *Northern powerhouse strategy.* London: HM Treasury.

HM Government. (2010) *Public Bodies Reform – Proposals for Change.* Available online: https://web.archive.org/web/20120112132526/http://www.direct.gov.uk/prod_consum_dg/groups/dg_digitalassets/%40dg/%40en/documents/digitalasset/dg_191543.pdf (Accessed 16th Sep 2017).

House of Commons. (2005). Office of the Deputy Prime Minister: Housing, Planning, Local Government and the Regions Committee HC 295–1. In *Empty Homes and Low–demand Pathfinders Eighth Report of Session 2004–05 Volume I Report, together with formal minutes, oral and written evidence.* London: The Stationery Office.

House of Commons. (2003). Office of the Deputy Prime Minister: Housing, Planning, Local Government and the Regions Committee HC 76–1. In *The Effectiveness of Government Regeneration Initiatives: Seventh report of Session 2002–03 Volume 1.* London: Stationary Office.

McHardy, A. (2001) "Moving tale of poverty". *Guardian Society,* July 4th p. 4.

Middlesbrough Council. (2005a) Executive report: Building sustainable communities in inner Middlesbrough. *Option Development Stage Report by Executive Member for Economic Regeneration and Culture and Director of Regeneration,* 19th April.

Middlesbrough Council. (2005b). *New vision for older housing: Issue 2.* Middlesbrough: Middlesbrough Council.

MORI, URBED, School for Policy Studies at the University of Bristol. (1999). *But would you live there? Shaping attitudes to urban living.* London: Urban Task Force.

Nathaniel Lichfield and Partners. (2004). *A housing vision for central Middlesbrough: Housing market assessment 2004. Report for Middlesbrough council.* Newcastle upon Tyne: Nathaniel Lichfield and Partners.

National Archives. (2011) Available online at: http://discovery.nationalarchives.gov.uk/SearchUI/image/Index/C11918866?source=20475 (Accessed 3 Feb 2014).

National Audit Office. (2007a). *Department for Communities and Local Government Housing market renewal: Report by the comptroller and auditor general | HC 20 session 2007–2008 | 9 November 2007.* London: Stationary Office.

National Audit Office. (2007b). *Homebuilding: Measuring construction performance.* London: National Audit Office.

Newcastle City Council. (2001). *Newcastle going for growth regeneration plan west end – Delivering an urban renaissance for Newcastle.* City of Newcastle upon Tyne: Newcastle City Council.

Newcastle Unison. (2002). *Our City is not for sale: Newcastle unison report.* Newcastle upon Tyne: Unison.

O'Brian, D., & Matthews, P. (2015). *After urban regeneration: Communities, policy and practice.* Bristol: Policy Press.

Office of the Deputy Prime Minister. (2005). *Delivering sustainable communities: The role of local authorities in the delivery of new quality housing.* Wetherby: Office of the Deputy Prime Minister.

Office of the Deputy Prime Minister. (2004a). *The Egan review: Skills for sustainable communities.* London: Office of the Deputy Prime Minister.

Office of the Deputy Prime Minister. (2004b). *Making it happen: The northern way.* London: Office of the Deputy Prime Minister.

Office of the Deputy Prime Minister (2003a) Statement by the deputy prime minister to the house of commons, Sustainable Communities: Building for the Future: Wednesday 5th February.

Office of the Deputy Prime Minister. (2003b). *Sustainable communities: Building for the future.* London: ODPM.

Office of the Deputy Prime Minister. (2002). *Community involvement: The roots of renaissance?* London: Office of the Deputy Prime Minister.

Polley, L. (2010). To hell, utopia and back again: Reflections on the urban landscape of Middlesbrough. In T. Faulkner, H. Berry, & J. Gregory (Eds.), *(2010) northern landscapes: Representations and realities of north-East England* (pp. 225–246). Woodbridge: Boydell.

Power, A., & Mumford, K. (1999). *The slow death of great cities? Urban abandonment or urban renaissance.* York: Joseph Rowntree Foundation.

Power, A., & Tunstall, R. (1995). *Swimming against the tide: Polarisation or progress on 20 unpopular council estates, 1980–1995.* York: Joseph Rowntree Foundation.

RIBA. (2015). *Design quality and performance: Initial report on call for evidence.* London: RIBA.

Robinson, F. (1997). *The City challenge experience: A review of the development and implementation of Newcastle City challenge.* Durham: Department of Sociology and Social Policy University of Durham.

Rodgers, R. (2017). *A place for all people: Life, architecture and social responsibility.* Edinburgh: Canongate.

Sayle, A. (2012) "How all politicians have made a mess of my beloved Merseyside" *The Observer,* 1st January p. 30.

Seyfang, G. (2010). Community action for sustainable housing: Building a low-carbon future. *Energy Policy, 38,* 7624–7633.

Shaw, K. (2000). Promoting urban renaissance in an English City: Going for growth in Newcastle upon Tyne. *Business Review North East, 12*(3), 18–30.

Smith, J., Blake, J., Grove-White, R., Kashefi, E., Madden, S., & Percy, S. (1999). Social learning and sustainable communities: An interim assessment of research into sustainable communities projects in the UK. *Local Environment, 4*(2), 195–207.

Sustainable Development Commission. (2005). *Sustainable buildings: The challenge of the existing stock: A technical working paper.* London: Sustainable Development Commission.

Townshend, T. (2006). From inner city to inner suburb? Addressing housing aspirations in low demand areas in NewcastleGateshead, UK. *Housing Studies, 21*(4), 501–521.

Urban Task Force. (1999). *Towards an urban renaissance.* London: E&FN Spon.

Waite, R. (2012). Shapps under fire over return to pathfinder. *Architects' Journal 14*[th], (June).

Walker, B. (2002). Unions blast Newcastle strategy. *Regeneration and Renewal 11*[th], (October).

Wates, N., & Knevitt, C. (1987). *Community architecture: How people are creating their own environment.* London: Penguin.

Weaver, M. (2005) "Market renewal homes sent back to the drawing board" *SocietyGuardian,* 23rd August.

Webb, D. (2010). Rethinking the role of Markets in Urban Renewal: The housing market renewal initiative in England. *Housing, Theory and Society, 27*(4), 313–331.

Wilks-Heeg, S. (2000). *Mainstreaming regeneration: A review of policy over the last thirty years.* London: Local Government Association.

Young, P., Dickinson, P. (2000) "The Battle for Scotswood: Residents pledge to fight plans for revamp". *Evening Chronicle,* Thursday June 8th 8–9.

Chapter 6
Reporting Corporate Sustainability: The Challenges of Organisational and Political Rhetoric

Christopher Gorse, John Sturges, Nafa Duwebi, and Mike Bates

The political voice of sustainability has broken. Presidential change in the USA has led to it hardening its tone, re-prioritising its stance and bringing a marked loss of climate change emphasis. The UK, another globally important economy, also amended its discourse making regulatory changes and dropping its commitment to zero-carbon buildings. The UK's impending departure from the European Union (BREXIT) brings future disconnection from EU directives, which had tied the country to reduced emission commitments.

Political rhetoric and regulatory change brings uncertainty to those industrial organisations that have aligned their strategies to take advantage of sustainable practice. These developments are too recent to predict the impact on the industry and corporate sustainability; nevertheless, the position that industry has adopted prior to these changes is interesting. Will those organisations already committed to sustainability continue maintaining a social or corporate interest or will the changes bring sustainable and economic uncertainty?

A review of annual reports of multinational corporations for years 2010–2012 was undertaken to understand the strategic positions that companies were taking with regard to sustainability. More recent reports and observations are then considered to provide insight into how organisations, both construction and nonconstruction, are reporting their current position. The review reflects on the corporate approaches conveyed to the outside world, the departures from their reports and the lack of certainty that current political changes are having on companies that have set their vision on sustainability investment. The review finds a divergence between the corporate sustainability strategies and the emerging rhetoric from some governments.

C. Gorse (✉) · J. Sturges · N. Duwebi
Leeds Beckett University, Leeds Sustainability Institute, Leeds, UK
e-mail: C.Gorse@leedsbeckett.ac.uk

M. Bates
PLProjects, Meltham, Holmfirth, UK

© Springer International Publishing AG, part of Springer Nature 2018   89
M. Dastbaz et al. (eds.), *Smart Futures, Challenges of Urbanisation, and Social Sustainability*, https://doi.org/10.1007/978-3-319-74549-7_6

6.1 Sustainability and Industrial Policy

In recent years, companies have been encouraged to improve their alignment with sustainability and social values (International Labour Office 2011). Evidence suggests that where companies trade on the open market, there are benefits in adopting sustainable practice. For example, there are differences in stock market performance between companies that are classed as 'low sustainability' and those classed as 'high sustainability' (Eccles et al. 2011). Similarly, Adams et al. (2014) noted performance differences when evaluating companies on the Dow Jones Sustainability Index between 2008 and 2009 and showed that those with low scores on the SAM index had poorer stock market performance and were less able to see and find shareholder value-creation opportunities. Eccles et al. (2011) reported that the benefits of those classed as 'high sustainability' companies were more strongly associated with the business to client sectors where companies competed on the basis of 'brand' and 'human capital' or were involved in extracting large amounts of natural resource. Early work also showed that companies adopting sustainable practice and committing to more environmental and social responsibility gain brand and reputation advantage, plus economic benefits in stock market value. However, many companies are reluctant to invest in sustainability, awaiting possible new legislation to support their endeavours.

Recently, most companies could be divided into two groups: those voluntarily adopting environmental and social policies and those maintaining a traditional model of maximising profits and responding to environmental concern only when required to by law and regulation (Eccles et al. 2011). However, the general message from governments and international offices then was that adopting sustainable practice would be beneficial and that those organising themselves for sustainable practice would be better positioned for the new markets coming along (International Labour Office 2011).

Upstill-Goddard et al. (2015) found that when gaining BES 6001 (BRE 2009) certification for responsible sourcing of products, the construction companies did not appear to adopt a strategic approach to certification, generally sticking with their existing environmental management systems rather than adapting to achieve greater recognition within the scheme. This is surprising, considering that companies may be competitively judged against the criteria.

In practice, for many companies tied to an ethical supply chain, organisational sustainability is now an established prerequisite, with prequalification questionnaires requiring environmental criteria to be met (Oke and Aigbavboa 2017; Hamani and Al-Hajj 2015). However, research on evaluating the value of strong environmental positions taken by companies when competing for work has found that the emphasis on sustainability was not a highly placed deciding factor. When evaluating tendering processes and ranking prequalification criteria, sustainability was not identified as a main or sub-criterion for bid evaluation and thus was not one of the factors listed that influenced success (Puri and Tiwari 2014).

Recently, it has been argued that sustainability is so important that it can no longer be ignored. Sustainability is integral to operation and survival and within stronger industry sectors is classed as innovative, forward thinking and capable of changing current markets. In some markets, it can be disruptive, creating a new way of working that displaces traditional practice. Innovation for sustainability is considered to be part of the 'sixth wave of innovation' and is something in which companies must engage if they are to be accepted by their peers and society (Silva and De Serio 2016). Different approaches to business operation, manufacturing and construction that are efficient, lean and have less environmental impact have a higher social value and can provide the brand image that sustains business activity and creates new markets; nevertheless, some of the major industries have appeared reluctant to engage.

Construction and contracting have been found to lag behind other sectors in the adoption of innovative practice (Peansupap and Walker 2004; Lamprinidi and Ringland 2008). Furthermore, the Global Reporting Initiative found that reporting the practice of sustainability was not well-established in construction when compared to other sectors, such as financial services or the electric utilities (Lamprinidi and Ringland 2008).

This chapter therefore seeks to examine the practices of heavy construction organisations and compare them with a sample of other multinational organisations, to interpret and better understand the corporate position companies are taking and their emphasis on sustainability. The discussion will also raise the question of how companies following a policy of sustainability may be affected by government rhetoric that places a different emphasis on future markets.

6.2 General Remarks on Sustainability

True sustainability is more than mere environmental impact mitigation; the aim is to leave the world as far as possible as we found it, for future generations' benefit. To do this, we must avoid polluting the atmosphere and the water supply, two items often taken for granted and which in the past have assumed to be 'free'. Early industry went in for such 'externalising' of costs, where they took what they required and left any mess for others to clear up and to bear the expense involved. To be sustainable, we should clear up our own pollution and not impose these costs on others who may lack the means to pay. Our impact on the planet can either be positive and a benefit or comes with a burden and price that others have to address.

We should also be sparing in our use of resources; Dieter Helm's words on natural and man-made capital are most apposite in this respect. Besides undesirable effluents and pollution, much manufacturing leads to the production of scrap or waste material as by-products. Improving processes to minimise scrap production and the recycling of such scrap are obviously desirable. We should also, wherever possible, source our raw materials from those classes of materials that are:

- Renewable
- Recyclable
- Available in such large amounts that the effects of their depletion are minimal

To avoid placing undue strain on resources and energy, it is desirable to source materials and labour as close as possible to manufacturing activities. Products should be designed to facilitate disassembly after use to enable either re-engineering or recycling. Good design involves tailoring materials to applications where their properties are appropriate. For example, using polymers, which are chemically and thermodynamically very stable for transient and ephemeral uses, is a misuse of materials derived from oil, a non-renewable resource.

6.3   Definitions of Sustainability

Based on information provided by the Institute for Sustainable Development, the concept of sustainable development was firstly referred to in the book *Silent Spring* published in 1962 when Rachel Carson described the interactions between the environment, economy and well-being, and the book is regarded as a turning point in our understanding of sustainable development.

Almost 8 years later, debate turned to discussion of the 'limits to growth'. The main question at that time centred on whether or not continuing economic growth would inevitably initiate dramatic environmental degradation and global societal breakdown. Reaching agreement on this subject was a complex task; however, following some years of debate and discussion, the resolution was reached that economic development could be sustained indefinitely but only if development is modified to take account of its ultimate dependence on the natural environment (Pezzey 1992a, b).

Although the essence of the concept of sustainable development is clear enough, its exact interpretation and definition has caused intense discussions. Ciegis (2004; 2008) suggests it is possible that the problem of terminology occurs in the concept's dual nature, covering development as well as sustainability; we must also be commensurate that sustainability is a term that the human race coined for their benefit, even if the aim appears laudable to preserve the plant and its environment in a 'natural state'. It is useful to attempt to understand the positions that are being taken when we refer to the term sustainability or sustainable development.

The literatures on sustainable development are mostly oriented towards separate sectors from which they are produced. It is worth noting that the concept of sustainable development is difficult to understand and define. Sustainability often has different meanings depending on the literature and the context in which it is used.

There are almost 50 definitions and circumscriptions of sustainability (Niels et al. 2005). A few examples are cited below:

World Bank (1987) Environment, growth and development. Development Committee Pamphlet 14, World Bank, Washington, DC

> ...satisfy the multiple criteria of sustainable growth, poverty alleviation, and sound environmental management.

> ...elevating concern about environmental matters...and developing the capacity to implement sound practices for environmental management...are [both] needed to reconcile, and, where appropriate, make trade-offs among the objectives of growth, poverty alleviation, and sound environmental management.

United Nations (2000). 'Cleaning Up Our Mining Act: A North-South Dialogue'

> Sustainable development: a concept that implies the precautionary principle. A healthy and continued multi stakeholder consultation will go a long way towards ensuring sustainability.

O'Riordan and Yaeger (1994)

> Sustainable development means adjusting economic growth to remain within bounds set by natural replenishable systems, subject to the scope for human ingenuity and adaptation via careful husbanding of critical resources and technological advance, coupled to the redistribution of resources and power in a manner that guarantees adequate conditions of liveability for all present and future generations.

United Nations Statistical Office (1992). 'SNA Draft Handbook on Integrated Environmental and Economic Accounting'

> Sustainable development means that economic activities should only be extended as far as the level of maintenance of manmade and natural capital will permit. A narrower definition of sustainability excludes the substitution between natural and man-made assets and requires maintenance of the level of natural assets as well as manmade assets. A sustainable development seems to necessitate especially a sufficient water supply, a sufficient level of land quality (prevention of soil erosion), protection of existing ecosystems (e.g. the virgin tropical forests) and maintaining air and water quality (prevention of degradation by residuals). In these cases, the sustainability concept should not only imply constancy of the natural assets as a whole (with some possibility of substitution) but constancy of each type of natural asset (e.g. of the specific ecosystems).

Although, Pezzey (1992a, b) argues that sustainability is simply regarded by analysts and modellers as a requirement to keep the consumption above some subsistence minimum, most other definitions understand sustainability as providing an improvement, or at least maintaining, in the quality of human life, rather than just sustaining its existence. Digging deeper into the concept is necessary to progress towards clarity of a definition. Yet, the determination of the essence of 'sustainability' cannot be an easy task and a concept to address all cannot be rigidly defined. O'Riordan (1988) and others have pointed out that the term is not always used correctly, and so definitions have been provided to clarify the meanings used.

The best-known definition of sustainable development was that of Brundtland (1987), and this report gave wide currency to the concept. However, our modern notions of sustainability date from the early 1970s. The UN Conference on the Human Environment in Stockholm in 1972, which addressed pollution issues in the developed world, emphasised that the environment was a vital development issue, while Third World countries were more concerned with poverty. The United Nations Environment Programme (UNEP) was created as a result of this conference.

Consequently, the concept of a 'sustainable society' first emerged in 1974 at an ecumenical study conference called by the World Council of Churches on Science and Technology for Human Development. The definition began by stating the conditions for social sustainability and the need for equity and democracy, rather than the more recent emphasis with environmental conditions. Interestingly, few people were aware of this work at the time, but it was the first time that concerns for global social and economic justice were given primary consideration. Previous conferences focused only on pollution and environmental issues and were perceived in the Third World as First World concerns and irrelevant to them. Only when the notion of equity and the social and economic dimensions were recognised could sustainability be seen as a truly international concern. So while the 1974 conference did not make a great initial impact, it was of primary importance and significance in the emergence of sustainability as we now understand it.

To further illustrate, the concept of 'sustainable development' surfaced in 1980 in the World Conservation Strategy published by the International Union for Conservation of Nature and Natural Resources. This document was written by a group of northern environmentalists and did not identify the political and economic conditions required to achieve sustainability. Because of these omissions, the document failed to place sustainable development firmly on the international agenda, and again, the document was seen as a First World luxury and not relevant to the Third World.

In 1983 the United Nations set up the World Commission on Environment and Development (WCED) under the Norwegian, Gro Harlem Brundtland. This commission produced the now famous report *Our Common Future* in 1987, which achieved two important things; it made clear that sustainability is about equity, both for current and future generations, and it gave sustainability international recognition and a high prominence.

The Brundtland Report drew together many important considerations that had previously been aired separately: the increase in population; the heavy demands likely to be placed on the Earth's resources of materials and energy, environmental impacts, etc.; and, most importantly, the need to change the direction of development towards achieving a more just and equitable treatment of every nation's needs while conserving resources for future generations' benefit. They called this 'sustainable development', famously defined as 'development which meets the needs of today without compromising the ability of future generations to meet their own needs'. The report addressed peoples of both the developed and developing worlds in saying that 'business as usual' was not a viable option, and because of this the report was universally accepted, the terms 'sustainability' and 'sustainable development' became common currency, and people have come to recognise its importance.

An important consequence of this was the so-called Earth Summit held by the United Nations in Rio de Janeiro in 1992. The objective was to turn Brundtland's broad aims into workable, sustainable policies by the governments of the world. The

Kyoto meeting followed in 1997, when the protocol on carbon emissions was drawn up and signed by some, but not all, nations. A second Earth Summit was held in New York in 2002, the 10th anniversary of the Rio meeting. These three conferences promised much, but delivered much less. They needed the co-operation of all major governments, and this did not happen. For example, the Kyoto Protocol was never signed by the USA or China, which saw a limit on carbon emissions as potentially damaging to their industries. Concerted universal action at government level has not been achieved, but work on measuring sustainability has continued in academia, in research institutes and in industry.

While Brundtland provided a comprehensible definition of sustainable development and raised its profile, the report offered no easy ways of measuring or achieving sustainable solutions to mankind's problems. However, over the 30 years since its publication, scientists, engineers, economists, sociologists and political thinkers have all worked on the problem. Recognition of the social and economic dimensions besides the environmental one led to the work of Elkington (1997) with his triple bottom line approach. Traditionally, economists considered the 'bottom line' in financial terms only, while Elkington made clear that there are social and environmental costs as well and that no sustainable solution could ignore them. The triple bottom line approach has been enshrined in most of the work done in the last 20 years on ways of measuring and assessing the sustainability of solutions to business, industrial and political problems. While international co-operation has been somewhat disappointing, methods of assessing sustainability have been developed. The implementation of the broad concepts set out by Brundtland has been thought through and developed in detail into forms that can be taken up by companies and organisations. In developing these assessment methods, it is not merely a matter of balancing the social, economic and environmental factors, but rather working out the details of a true synthesis. One example of this work is that of Epstein and Roy (2003) who identified the following nine principles for improving sustainable performance:

1. Ethics: organisations should establish, promote, monitor and maintain ethical standards in the way they deal with the organisation's stakeholders.
2. Governance: this is a commitment to manage all resources effectively, recognising the manager's duty to focus on the interests of the organisation's stakeholders.
3. Transparency: the organisation provides timely disclosure of information about its products, services and activities, thus permitting stakeholders to make informed decisions.
4. Business relationships: the organisation engages in fair-trading practices with suppliers, distributors and partners.
5. Financial return: the organisation compensates providers of capital with a competitive return on investment and protection of organisation's assets.
6. Community involvement/economic development: the organisation fosters a mutually beneficial relationship between the organisation and community in which it is sensitive to the culture, context and needs of the community.

7. Value of products and services: the organisation respects the needs, desires and rights of its customers and strives to provide the highest levels of product and service value.
8. Employment practices: the organisation engages in human resource management practices that promote personal and professional employee development, diversity and empowerment.
9. Protection of the environment: the organisation strives to protect and restore the environment and promote sustainable development with products, processes, services and other activities.

Another example is due to Silvius and Schipper (2012), who identified various key principles derived from the concepts of sustainability, as follows:

- Sustainability is about balancing or harmonising social, environmental and economic interests.
- Sustainability is about both long- and short-term considerations.
- Sustainability is about local and global orientation.
- Sustainability involves consuming income not capital.
- Sustainability involves transparency and accountability.
- Sustainability is about personal values and ethics.

6.4   A Brief History of Industrialisation, Sustainability and Governance

The term sustainability attained prominence with the publication of the Brundtland Report in 1987. Since then, various authors, including Elkington (1999), Dresner (2008), Caradonna (2014) and Washington (2015), have reviewed its development. In tracing the origins of the ideas that led to our present conception, we can go back to the sixteenth century. Five hundred years ago, probably the most versatile material we possessed was timber, as it was used for the construction of buildings, of ships, as fuel and as a reducing agent for the smelting of iron. The countries of Northern Europe had great forests, and supplies of timber were readily available. However, as societies developed and populations expanded, demand for timber also grew and the rate of consumption increased, to the point where people became alarmed that their great forest areas were disappearing. Forests are highly visible assets, and their loss was noticed. While the concept of sustainability had not yet emerged, it was recognised that conservation steps were required to preserve forests; otherwise they would be lost, perhaps forever. Responsible people knew that mature forests could take a century or more to regenerate, and because of this, the earliest writings and legislation for the protection of woodlands date from the sixteenth century. In England, parliament passed the Act for the Preservation of Woods (1543) (Moorhouse 2005). The opening words of the Act make this clear:

The King our Sovereign perceiving and right well knowing the great decay of timber and woods universally within this his realm of England to be such, that unless speedy remedy in that behalf be provided, there is great and manifest likelihood of scarcity and lack, as well of timber for building, making, repairing and maintaining of houses and ships, and also for fuel and firewood for the necessary relief of the whole commonality of this his said realm.

In Germany, the earliest writings on sylviculture date from 1713 (von Carlowitz 1713). In Japan Tokugawa Ieyasu founded the Tokugawa Shogunate in 1603, effectively unifying Japan, ending decades of internecine warfare and ushering in a long period of peace and prosperity. This led to a great increase in building construction, sharply increasing the demand for timber. The threat of deforestation loomed, and the Shogun's government adopted enlightened steps to remedy the situation (Diamond 2005). Their policy was implemented in three stages. Firstly, logged land was closed off to allow forests to regenerate. Secondly, guard posts were established on all roads and rivers to check on people shipping timber. These guards monitored the woodlands to ensure that the rules were obeyed. Finally, the government specified and updated the rules. The Shogun controlled a quarter of Japan's forests, and he issued the rules about timber use. These measures proved to be very effective. It has been pointed out that historically, many societies that cut down all of their trees were unable to survive (Diamond 2005); thus governance over our natural resources not only impacts on economy but on the survival of communities.

China began its headlong drive to industrialise in the 1980s. This began slowly, but then the pace rapidly picked up so that timber consumption increased sharply and the threat of deforestation began to appear. In the mid-1990s the Chinese government issued an edict limiting the use of local timber to prevent deforestation occurring.

The next contribution to the sustainability concept was that of Malthus (1798); his famous theory was published in 1798 about looming mass starvation. He was a clergyman who had studied crop yields and population statistics, and he predicted that the human population would eventually be limited by starvation, unless more food could be produced. In his review of the development and evolution of the principles of sustainability, Dresner (2008) identified the work of Malthus over 200 years ago as the beginning of the appearance of the term. His review outlines the relative importance of all the factors involved including population, consumption and depletion of resources, environmental impacts as well as the social and economic dimensions.

However, during the nineteenth century, Spanish colonists in South America discovered the existence of huge cave systems in Peru, which had provided nesting sites for millions of bats and birds over tens of thousands of years. These caves were filled to a great depth by very nitrogen-rich droppings. This guano made excellent fertiliser, and a large and growing fleet of ships began plying between South America and Europe bringing large cargos of guano for use as fertiliser. Many fortunes were made from this trade, and as a result, crop yields improved and the threat of starvation receded.

By the end of the nineteenth century, people became nervous about what would happen when guano supplies were exhausted, and this concern motivated a German chemist called Fritz Haber to develop a laboratory-scale process to synthesise ammonia (NH_3) from hydrogen and nitrogen. Nitrogen exists as a diatomic molecule (N_2) in the Earth's atmosphere in large quantities, composing 79% of the Earth's atmosphere. Splitting this molecule is an extremely endothermic process, and Haber had to employ a combination of high temperature and high pressure to do it. He then teamed up with a German chemical engineer called Robert Bosch, and together they developed an industrial-scale process to manufacture ammonia in tonnage quantities. This process came on stream in 1913, but its wide-scale exploitation only occurred after the Second World War. Highly nitrogenous, artificial fertilisers have been used in Europe, in North America and in China, and in fact, these fertilisers have been over-used to the point where nitrogen pollution has become a major problem.

In the late nineteenth century, the Swedish chemist Arrhenius predicted that with industrialisation, the CO_2 content of the Earth's atmosphere would increase and warming would occur. This has happened; in 1900, the CO_2 content of the air was around 275 parts per million (p.p.m.), whereas today it stands at over 400 p.p.m., a huge increase and the driver for current global warming. Studies of ice cores have traced how the CO_2 content of the atmosphere has changed over many thousands of years, changes that have been correlated with past temperature regimes.

The problems of atmospheric, water and soil pollution were identified as industrialisation developed during the late nineteenth and early twentieth centuries. The winter 'smog' problem was recognised and correctly ascribed to uncontrolled smoke emissions from houses and from industry. Fish disappeared from rivers due to uncontrolled release of industrial effluents into rivers and streams. The construction of buildings on 'brown field' sites was inhibited because of pollution by toxic species in the soils of former industrial areas. Appropriate legislation was enacted and enforced by factory inspectors and environmental officers from local councils, and these problems have been ameliorated in the UK.

As the scale of human impact has grown, so has the realisation that these factors are all interlinked and are all connected to the growth in the Earth's human population. Much effort has been devoted to both the measurement and assessment of these impacts, but their scale has continued to grow over the years since publication of the Brundtland Report. Indeed, some people have become so concerned that they have even questioned whether *Homo sapiens* can survive beyond the twenty-first century (Rees 2003).

A new approach to sustainability has been proposed by Rockström et al. (2009); his research group has identified human impact on a global scale in nine key areas, and these are:

• Biodiversity loss
• Climate change
• Nitrogen misuse
• Land use

- Fresh water supplies
- Toxic species release
- Aerosol particles
- Ocean acidification
- Damage to ozone layer

From the global picture, they have identified those areas where urgent action is required, many of these areas being interlinked. For example, land use impacts upon biodiversity. The trend in agriculture has been to create ever-larger fields so that farmers can take advantage of the latest agricultural machinery. However, removal of hedgerows has reduced biodiversity by destroying the habitats of numerous insects, small creatures and birds. Most of these creatures had a beneficial effect in controlling pests, and farmers now use pesticides as well as fertilisers. Heavy machinery can cause soil compaction, which leads in turn to greater soil erosion and run-off of fine soil particles and nitrogen during rainfall. So in this one example, land use, biodiversity and nitrogen pollution are interlinked.

It may seem surprising that biodiversity should be placed so high in this list, but we now recognise that we are part of the planet's ecosystem and do not control it. The mark of a healthy ecosystem is that it contains a large number of species, and if we lose too many, the ecosystems on which we depend will not be able to cope with the demands that we place upon them. We are only now learning to recognise which species can signal danger if they are lost. There is a sense in which the CO_2/climate change lobby (important as it is) have captured the sustainability topic to the exclusion of other factors. The Rockström approach is therefore very useful in reminding of the other factors that are also important. More recently, the International Institute for Sustainable Development (IISD) outlined the timeline of sustainable development as shown (Tables 6.1 and 6.2):

These reports and events are of major media and political interest but without recognition within the major industry sectors little will change. It is important to understand how sustainability as a company attribute is being adopted by different organisations and companies.

6.5 General Observations from Company Annual Sustainability Reports

A review of a number of company annual sustainability reports, from different sectors, offers a brief insight into how changes are being made to shape operating procedures to make organisations less environmentally damaging, more sustainable and socially acceptable. The extracts provide a flavour of how different the 'on-the-ground' approaches are. A few examples from different sectors are provided in the section below:

Table 6.1 Sustainable development timeline

Sustainable development timeline
1962 *Silent Spring*
1967 Environmental Defence Fund (EDF)
1969 National Environmental Policy Act
1970 First Earth Day
1971 International Institute for Environment and Development (IIED)
1972 UN Conference on the Human Environment and UNEP
1977 UN Conference on Desertification
1980 Independent Commission on International Development Issues
1980 Global 2000 Report
1982 The UN World Charter for Nature
1985 Climate Change
1987 Montreal Protocol on Substances that Deplete the Ozone Layer
1987 Our Common Future (Brundtland Report)
1988 Intergovernmental Panel on Climate Change (IPCC)
1989 Stockholm Environment Institute
1990 International Institute for Sustainable Development (IISD)
1990 Regional Environmental Centre for Central and Eastern Europe
1990 UN Summit for Children
1991 Global Environment Facility
1992 The Business Council for Sustainable Development
1992 Earth Summit
1993 First meeting of the UN Commission on Sustainable Development
1996 ISO 14001
1999 Launch of the Dow Jones Sustainability Indexes
2000 UN Millennium Development Goals
2002 World Summit on Sustainable Development
2002 Global Reporting Initiative
2005 Kyoto Protocol enters into force
2005 Wal-Mart institutes global sustainability strategy
2010 The Economics of Ecosystems and Biodiversity final report
2012 One of the first of the Millennium Development Goal targets is achieved
2012 Rio +20
2015 Paris Agreement UNFCCC

Duwebi (2017)

Table 6.2 Heavy construction sector

Company in the SAM index	Number of reports found containing relevant references to sustainability
Acciona	2 (3)
Fomento de Construcciones y Contratas	3 (4)
GS Engineering and Construction	1 (1)
ASC	0 (0)
Hyundai E&C	0 (0)

6.6 Automotive, Mining Industry, Mobile Telephone Industry and Banking

Daimler, in the automotive sector, recently issued a directive to make the utilisation of as much recycled material as possible mandatory in the manufacture of their new vehicles. Since Daimler is highly regarded as a quality brand in this market, this commitment sends a powerful signal regarding the use of recycled materials to others in the sector.

The automotive sector has suffered recent emission scandals; sustainability commitments beyond emissions will assume greater importance as it attempts to rebuild its reputation. Once the statement goes public, social discourse through the web follows, with companies being held socially and publically accountable through the media. It will be interesting to see how such statements are measured in future years; nevertheless, a practical commitment has been made by Daimler and others in the industry.

Traditionally, mining has not had a good environmental image. However, most global mining organisations are now taking action to improve their reputation and bring positive impact to local economies. Formerly, once the minerals, ores and other materials were extracted, mine workings were usually abandoned leaving unsightly and often hazardous conditions. These companies now need to show more social responsibility, and their practices are changing. In their 2012 sustainability report, Anglo-American published and cited practice to be adopted when a new deposit of minerals or ores is identified for exploitation. The new site is surveyed in detail, and a full inventory of all the flora, fauna and natural features is recorded. Where necessary, samples of the plants are taken for cultivation and storage so that the site can be returned to its pristine state once mining operations have ceased. Clearly, while the geological impact is felt, the natural fauna, flora and wildlife are preserved. This in itself is a step forward, but Anglo-American also demonstrate good social awareness, recognising that a healthy and well-educated local workforce will benefit them. To this end, they help to fund local infrastructure, health clinics, schools, housing and other amenities necessary to create a community that becomes mutually beneficial from the mining activity. Where roads are constructed as part of mining and processing needs, the road is built to a sufficient standard to form part of the local infrastructure once mining operations have ceased.

Industries are much more aware of the waste of resources and are taking steps to curtail and recycle such waste. Subtle changes to mass-production processes can have a major impact. China Mobile, of the telecommunications sector, issued a directive to reduce the amount of plastic waste in the manufacture, processing and packaging of their products. A modern mobile phone handset will weigh perhaps 100 grammes, but as production is large, into the millions, the waste produced and plastic used are substantial. 10,000 handsets will weigh perhaps 1 tonne. In a recent report, China Mobile claimed to have improved manufacturing measures and achieved a reduction in plastic waste of around 8000 tonnes per annum. The adoption of such measures has a considerable impact.

It is interesting that changes to business processes that affect the way the consumer or customer interfaces with the supplier of a product or service can have a huge impact on the energy consumed and the travel needs required to access the service. The Itau Bank of Brazil recently provided details of how they contribute to sustainability by introducing operating systems designed to reduce the number of journeys that customers need to make to their local bank using their personal banking systems. In their investment banking systems, they have an investment appraisal set-up which analyses requests for investment funds in terms of sustainability and environmental impact, and those proposals deemed to be environmentally damaging will not be funded. Effective weeding out of such proposals represents a major contribution to improving sustainability.

6.7   Adoption of the Sustainability Term in Industry

In the late 2000s, increasing attention was given to sustainability by industries in the USA and EU, and progress towards more sustainable development rests on the balanced achievement of economic development, social advancement and environmental protection. Industry's main concern at that time was achieving green economy, green industry and green growth to help achieve sustainable development and move consumption and production into a more sustainable path in the long term.

Green industry mainly focused on the industrial production and development that promotes environmental, climate and social consideration into operations. Therefore, the main aim is about greening industry through a continuous improvement of environmental performance regardless of sector type or locations. United Nations Industrial Development Organization (UNIDO) also referred to the need for achieving green industry by reducing the environmental impact of processes and products throughout:

• Improving production efficiency: using resources more efficiently and optimising the productive use of natural resources
• Enhancing environmental performance: minimising environmental impact by reducing waste generation, emissions and environmentally sound management of residual wastes
• Minimising health risks: caused by environmental emissions, along with the provision of goods and services that minimise the occurrence of such emissions

Another helpful element is green technology. Promoting best environmentally sound technologies helps in improving the production of goods that reduce negative environmental impact. A green industry technology recognised by UNIDO is 'one which is incorporated or woven into the economic, social and environmental structures and best serves the interests of the community, country or region that employs it'. Some examples of green technology can be illustrated in the creation of tools and mechanisms to improve water productivity technology in order to reduce water consumption, also those innovative technologies to increase energy efficiency.

Moreover, encouraging companies from different sectors to take more responsibility for their operation and participate in so-called design for environment (DfE) was found to be helpful in producing more durable, reusable and recyclable goods, products and services.

Very soon after deciding to explore how large corporations integrated sustainability principles into their management, it was found that the leading companies published annual company reports which included substantial sections dealing specifically with sustainability. Many of these companies also published separate annual sustainability reports, all of which were downloadable. Much can be gained from company reports, not least a public commitment by the company to practise and change. It has been recognised that these reports can potentially form the basis of a study and continuous reflection. This initial insight is just that, but over time, reflections on how the content of these reports change and how practice relates to the reports will become of increasing public and political interest. The global corporate giants have huge influence, but equally they are exposed to public criticism which can affect the markets they serve.

6.8   Product Life Cycle and Unexpected Consequences

Nowadays, it is usually necessary to consider the product life cycle in judging whether a product and process is sustainable. The second half of the twentieth century was occupied with cleaning up the results of early industry's failure to consider the fate of its products once they left the factory gates and were sold. Victorian manufacturers considered that their responsibilities ended with the sale of their products. Nowadays, manufacturers are increasingly compelled to consider their products' end-of-life scenarios, in other words, 'cradle to cradle' approach to consider the whole 'extended life cycle'.

It is important to consider both the process and the product when judging whether something is sustainable. For example, the product may be deemed to be a good contribution to the achievement of sustainability, but the process by which it is produced may not be or indeed be harmful. An example of this is low-energy light bulb manufacture, promoted in the *Code for Sustainable Homes*. Such light bulbs are filled with mercury vapour and are manufactured in China where health and safety legislation has been minimal or non-existent. People living in the city around the plant where the bulbs are made are suffering from major health problems caused by mercuric poisoning. The light bulbs are used in Western Europe in huge numbers, while the people of China pay a high price in ill health; therefore this process cannot be regarded as truly sustainable. A more holistic approach is needed. Governments can have a great influence on the markets that develop; recently we have seen some changes that are challenging sustainability and the direction that companies may be willing to take.

6.9   Sustainability: Recent Political Developments and Reactions

A degree of political uncertainty in energy and sustainability policy has been brought about as a result of recent political developments such as BREXIT and the US election (Watson 2017). Companies are concerned that their markets will be affected, and leading firms, including Kingfisher, BAM and ARUP, have lobbied the UK government to use its Clean Growth Plan and other political tools to tackle emissions from buildings (letter to Greg Clark, Secretary of State, WWF 2017).

Concern has been raised by the head of energy and climate change at WWF:

> The low carbon economy represents a huge opportunity for UK businesses so it's no wonder that they're desperately looking for longer term clarity that will enable them to invest in the technologies that we know can help to tackle climate change…. (Bairstow 2017)

A change in direction of the USA was signalled recently as the president signed an executive order placing economic and employment concerns above those of the climate change agenda (Merica 2017). President Trump argued that both growth in business and tackling issues of climate change can sit alongside each other, which is not so dissimilar to the position set out by the UK's Chancellor of the Exchequer in its 'Fixing the foundations' review.

> Productivity is the challenge of our time. It is what makes nations stronger, and families richer. Growth comes either from more employment, or higher productivity… …we need to focus on world-beating productivity, to drive the next phase of our growth and raise living standards. (Osborne 2015 p.3)

There is evidence that the UK Department for International Trade is prioritising trade and growth ahead of climate change and illegal wildlife trade, in the wake of the BREXIT vote (Shipman 2017). In both the USA and UK, this discourse signals a relaxation of some the climate change legislation in favour of trade and economic growth.

As a result of this discussion, the question arises: if companies have positioned themselves for sustainability, what impact will the political uncertainty have on their market value? As these political changes are so recent, this research will not be able to expose evidence to reliably address this; however, the insights into company activity through their policies will provide useful context to explore the investments that companies have made. As time unfolds, future study will be able to explore how such events impact on an organisation position and how the statements which organisation make change in response to political positions. Notwithstanding these recent changes, it is useful to look back at the positions that companies have taken and why there is so much concern over recent political statements.

6.10   Analysis of Company Reports: Research Method

The work presented here represents part of a recently completed study (Duwebi 2017), which reviewed multinational global company reports and also a reflection in the light of the recent political changes to the sustainability agenda. The review exposes the position companies were taking and the strategy that they were adopting during the period 2010–2012.

In this research, ten industry sectors, based on the Dow Jones Sustainability and Robeco SAM Sustainability Index, were reviewed. In each sector, the five highest-ranking companies identified from the SAM index were selected and related reports captured and the content assessed. While this study limits the research to those companies considered most sustainable by the index, it is of interest to know the sustainability positions these companies are reporting and how their experiences compare. For any organisation, future events can unfold as predicted, but also, and, just as importantly, in unpredictable ways. It is of interest to reflect on the sustainable positions adopted in the reports and the events that have unfolded.

6.11   Sample Considered

The ten sectors considered for this research project were:

Aerospace and defence
Automobile
Banks
Chemical
Food and drug retailers
Heavy construction industry
Mining industry
Mobile and telecommunications
Oil and gas producer
Pharmaceutical

However, this section will just consider three sectors, the heavy construction sector, with the mobile and chemical sectors for comparison purposes. It will examine how the various company's interpretations of sustainability align with the Brundtland definition and also how they have positioned themselves in taking up the sustainability agenda, prior to the most recent political developments.

In selecting the sample, purposive sampling was adopted to capture an insight into the sustainability rhetoric of the higher ranking organisations. The review of archival research data provides insight into that reported and also provides a baseline against which any deviation between planned and actual activity can be interrogated.

The work is cross-sectional in that each report provides a snapshot of the company activities at that moment in time and across companies and is also longitudinal in that annual reports for 3 years 2010–2012 were investigated (Duwebi 2017). In light of recent political changes and to capture any trends or departures, more recent reports were discussed, but not investigated to the same degree as the main body of data reviewed. Threads of enquiry were followed, with the discussion following points of related interest. The more recent evaluation must be considered as an insight and prompt for further research rather than systematic enquiry.

Different types of company reports were recognised as information sources for this study and included risk management reports, citizenship reports, corporate social responsibility (CSR) reports, sustainable development (SD) reports, corporate reports (CR), beyond the mine (BTM) reports and summary review reports. Overall, 308 reports were included for review.

6.12 Industrial Definitions of Sustainability: Alignment with Brundtland

With the aid of NVivo software, queries relative to the facet were used to explore and extract related data. It was recognised that many of the company reports were not produced with sustainability or the management of sustainability in mind. The method of searching and querying was iterative, looking for terms, capturing information, recognising different related terms used (compared with those initially searched) and then repeating the search, using broader search terms. In this review, we stick to those terms which address sustainability and do not drill down to the level of detail beyond sustainability terms that were reported by Duwebi (2017).

6.13 Discussion: Sustainability Within Company Reports

Analysis of the raw data relating to use of the term 'sustainability' showed that its use is not limited to environmental and humanitarian definitions, as would be expected. Thus, the context within which the term is used is fundamental to its interpretation. The information surrounding the word sustainability was expected to be of more relevance than the term itself. Considering the context of individual words is a logical step in developing understanding, and, in most cases, it becomes relatively easy to distinguish the nature of a word's meaning and intent as the surrounding discourse is considered. Thus once the term sustainability is used within any document, it is important to consider the context and how companies articulate the position into their operation and processes. The occurrence of discourse related to sustainability was noted and recorded in the tables below (Tables 6.2, 6.3 and 6.4).

Table 6.3 Automobile sector

Company in the SAM index	Number of reports found containing relevant references to sustainability
BMW	5 (9)
Daimler	6 (9)
Fiat	1(1)
VW	1 (1)
Toyota	0 (0)

Table 6.4 Chemical sector

Company in the SAM index	Number of reports found containing relevant references to sustainability
AkzoNobel	1 (1)
BASF	1 (1)
Bayer	6 (9)
DMS	3 (17)
Dow chemical	0 (0)

Not all heavy construction companies are making commitments or use of the term sustainability. At present, the leading organisations in this sector are not seen to be lagging behind those in other sectors. In the larger sample of research, industries such as aerospace, banking, food and drugs, mobile sector, oil sector, automobile, chemical, mining and pharmaceutical were reviewed. In comparison with these, the degree to which the heavy construction sector use 'sustainability' rhetoric within their reports is comparable to that used in mining and pharmaceuticals. It is interesting that those companies using a greater amount of sustainability discourse were the chemical and automobile sectors. Within these sectors, the top listed companies were most explicit about their commitments, placing greater emphasis on how they are progressing in this respect.

BMW's (annual report 2010, p.171) dialogue aligns closely with the philosophical content of the Brundtland definition, giving equal consideration to ecological, social and economic development. They also use the term sustainability from an organisational perspective, referring to the relevance of corporate sustainability and its importance in three areas: resources, reputation and risk.

Similarly, Daimler (Daimler Annual Report 2010, p.251) defines sustainability in line with Brundtland, as:

> Using natural resources in such a way that they continue to be available to fulfil the needs of future generations. In the view of the Daimler Group, sustainable business operations have to give due consideration to economic, ecological and social aspects.

Also, the following year's Daimler Sustainability Report (2011, p.10) defines sustainability as:

> Responsible corporate behaviour that leads to long-term business success and is in harmony with society and the environment. The company moves toward its goals by making sustainability a firmly integrated aspect of their operations and by requiring and promoting a

strong sense of responsibility for sustainable operations among all of their managers and employees throughout the Group. They include their business partners in this process and participate in continuous dialogue on these issues with their stakeholders.

With a lesser degree of emphasis than the previous two, Volkswagen's report (2012a, b) (p.16) interprets sustainability as:

A call for a balance of economic, environmental and social objectives.

The comments from the Volkswagen group preceded the emission violations, which started to emerge following a 'tip-off' in 2014. However, it is claimed that some of the cars affected were in production as early as 2009 (Atiyen 2017). A recent review of the affair concluded that the root cause of the 'unethical scandal goes back to business culture and structure of the company'. This suggests that compliance-based business ethics, like those used by Volkswagen, are failing to treat employees ethically and present employees with the dilemma of either losing their jobs or taking unethical action. The difference between the report and actual company culture and practice that led to the problem is evidenced through the emission scandal and company confession. Volkswagen attracted all the opprobrium for this, but they were following EU-wide policy in the drive to reduce carbon dioxide emissions. Government policy was based on the single issue of carbon dioxide to the exclusion of all else, including air quality and health considerations. Other manufacturers have since owned up to similar practices, and this incident points up the dangers inherent in maintaining a single-issue focus by governments.

Returning to company reports from other automotive companies, the Fiat Group believes that:

Sustainability is not an objective in and of itself, but rather a journey of continuous improvement essential for long-term growth. (Fiat S.P.A. Sustainability Report (2011) p.12)

Fiat makes no specific reference to the three aspects of the triple bottom line in any one part of their reports; instead, they refer to the three elements either independently, or as shown below, with social and environmental aspects being considered.

Over the years, the Group's sustainability strategy has resulted in a variety of projects designed to promote increasingly sustainable mobility, help protect the environment and natural resources, ensure the health and safety of employees, invest in their professional development, and build a constructive relationship with local communities and commercial partners. The desire to continue growing in harmony with people and the environment is embodied in the Sustainability Plan. (Fiat S.P.A. Sustainability Report (2011) p.12)

The comment places emphasis on a culture of growing professional development.

This sector also provided the highest number of references to sustainability.

In their annual report 2010 (p.205), BASF Corporation aligns to the Brundtland definition of sustainable development. Meanwhile, the AkzoNobel 2012 report (p.163) only makes an oblique reference to global sustainability. They see sustainability as connected to every area of business, stating that:

By doing more with less, sustainability value will be fundamentally connected to business value. We are making sustainability profitable by tailoring solutions to our customers' needs today and in the future and by future-proofing our supply chain.

Another chemical corporation DSM (Dutch State Mines) introduced their approach and understanding of sustainability in their integrated annual reports 2010 as a way of creating brighter lives for people today and in future through connecting their unique competencies in life and material sciences to create solutions that nourish, protect and improve performance. DSM focuses on the triple bottom line of economic performance, environmental quality and social responsibility pursued simultaneously to create value for all stakeholders. Furthermore, in the same report (p.11), they state why the commitment is important for the company's future: 'sustainability will be the key differentiator and value driver over the coming decades'.

In similar vein, the Bayer 2010 sustainability report (p.6) refers to the importance of creating a balance between the three aspects of sustainability by stating that 'we can only be successful in the long term if we take ecological and social needs into account as well as economic considerations'.

All of the positional statements commit to sustainable development to ensure readiness for future markets.

6.14   Heavy Construction

Within the companies listed, the evidence suggests that the heavy construction sector is as cognisant of sustainability as other sectors and is concerned with improving its social, economic and environmental sustainability indicators.

Acciona's 2010 annual report (p. 10) referred to the company's commitment to sustainability as going beyond generating economic value and stated that 'we aim to contribute to development with a balanced business model for the benefit of future generations'. They then suggest that the company's sustainability master plan 2010–2013 (annual report 2010) rests on six pillars: innovation, environment, engagement with society, people, the value chain and governance, all aimed at achieving concrete goals of sustainability.

Fomento de Construcciones Contratas (annual report 2012) refers to the importance of sustainability in both construction product and process. This report (p. 447) states that 'sustainable construction refers not only to managing environmental impact while the works are being executed, but also to management of the "product" throughout its useful life'.

GS Engineering and Construction (annual integrated report 2012, p. 6) differs from this view and instead interprets sustainability as 'creating value that can be shared among various stakeholders, as well as fulfilling their responsibilities as a corporate citizen'. Their vision of sustainability (annual integrated report, p. 7) relies on 'maximising organisational competence based on core values of great innovation, great challenges and great partnerships to earn trust to grow as a sustainable global company'. The aim is defined as 'Pursue Growth by Creating Sustainable Value Together' (Annual integrated report 2012).

The research shows that the use of the term, related terms and expressions, linked to sustainability (environmental, social and economic) creates company principles that add benefit when considering future markets.

6.15   Later Information: Acciona – A More Recent Enquiry

A review of Acciona annual report in 2015a found no reference to 'sustainability', but there is a significant shift from earlier reports. The growth in renewable energy and the need for strategic positioning with emergent markets, which need to be sustainable, were recognised as a key business driver. The company has shifted from an acknowledgement of sustainability issues to one where it has positioned itself to take advantage of emergent economies and the need to engage with sustainable energy supply and also to adopt sustainable practice. Their sustainability report (2015b) complements their annual report. It details training and development initiatives – performance incentives linked to sustainability – and covers many aspects of practice and makes strong commitments. Interestingly, this was done even after the preceding years were seen as financially challenging and where the company had to restructure itself. Sustainability and business development are no longer seen in opposition but very much one and the same.

> Acciona has assumed these challenges (United Nations General Assembly Sustainable Development Goals) as its own, and incorporated them into its business model... ... (with a plan) to make Acciona a carbon neutral company. (Acciona Sustainability Report 2015b)

A key value remains a 'concern for the environment':

> ACCIONA sees the fight against climate change, sustainable use of natural resources and protection of biodiversity as the main principles of its environmental strategy. Acciona (2015a), p.13

> To play a leading role in transforming the planet's infrastructure and in sustainable energy, while focusing on having a strong balance sheet, remunerating our shareholders appropriately, and constantly seeking growth opportunities. Acciona Annual (2015a), p.10

> In this line, last September, the United Nations approved the 17 sustainable development goals that will guide the investment agenda for governments, multilateral funds and private investors towards emerging economies in order to outfit them with infrastructure that contributes to their development.

> In this context. The integration of our construction, service and water businesses into a single Infrastructure division and the steps taken to enhance the selection of business opportunities, as well as risk and contract management, have begun to produce results, since order intake increased by 22% and our backlog by 13% in 2015. Acciona (2015b), p.9

The reports and discussion show deeper commitments being made to a sustainable agenda and are interesting in that where companies like VW diverge from these, market and legal consequences follow. We now witness political changes that the companies will be responding to in future. The reports and positions taken by the organisations over the next decade, in response to these changes, will be interesting.

6.16   Conclusion

The reports show that international corporations are making reference to, adopting and in some organisations embracing the concept of sustainability including its environmental, social and economic dimensions.

Over the years, many have positioned themselves for sustainability and the emergence of related markets. The review of Acciona's more recent company reports in 2015 shows that the company has aligned the whole operation to embrace sustainability and makes the company ready for the emergence of the new energy and sustainability markets. Some remain cynical that such statements are still part of relationship propaganda; however, such public commitments expose the organisation and make them publically or socially accountable.

Reports for 2016 are not yet available, so the evidence of how companies will react to recent political changes and relaxation of legislation cannot be reported. However, at a UK domestic level, there is some evidence of government lobbying, by construction firms, requesting clarity and change for the industry in the form of the Clean Growth Plan. It is clear that over the years, commerce has acknowledged the need to develop sustainable policies and to invest to be ready to engage. The lack of clarity is concerning for those in the industry that have invested to ensure that they can continue to operate when sustainability legislation is introduced, whatever effect the government policy changes will have when the company's sustainability investments have potential to mature.

The indications from recent political changes show that governments are prioritising economic growth and trade ahead of climate change and environmental protection measures.

References

Acciona (2015a). Annual report 2015, available http://annualreport2015.acciona.com/d/annual-report-2015.pdf

Acciona (2015b). Sustainability report, available http://annualreport2015.acciona.com/d/sustainability-report.pdf#_ga=1.194447213.1788239440.1490796301

Adams, M., Thornton, B., Seperhri, M. (2014) The impact of the pursuit of sustainability on the financial performance of the firm. *Journal of Sustainability and Green Business*, 2. http://www.aabri.com/manuscripts/10706.pdf

Atiyen, C. (2017). Everything you need to know about the VW diesel-emissions scandal. Car and Driver, 10th March 2017. Available http://blog.caranddriver.com/everything-you-need-to-know-about-the-vw-diesel-emissions-scandal/

Bairstow, J. (2017). Firm call on government to tackle building emissions. Enery Live News, Energy Efficiency. Available http://www.energylivenews.com/2017/03/28/firms-call-on-government-to-tackle-building-emissions/

BRE. (2009). *Certification scheme for responsible sourcing of construction products*, SD186, rev 2, 20 July 2009. Available https://www.bre.co.uk/filelibrary/greenguide/PDF/SD186_Rev2_ResponsibleSourcing_SchemeDocument.pdf

Bruntland. (1987). *Our common future - Brundtland report*. Oxford University Press: United Nations. http://www.un-documents.net/our-common-future.pdf

Caradonna, J. L. (2014). *Sustainability. A history*. Oxford: Oxford University Press.

von Carlowitz, H. C. (1713). *Sylvicultura oeconomica*. Leipzig: Oder haußwirthliche Nachricht und Naturmäßige Anweisung zur wilden Baum-Zucht.

Ciegis, R. (2004). *Ekonomika ir aplinka: Subalansuotos plėtros valdymas*. Kaunas: Vytauto Didžiojo universiteto leidykla.

Ciegis, R., & Ciegis, R. (2008). Laws of thermodynamics and sustainability of economics. *Inzinerine Ekonomika-Engineering Economics, 2*, 15–22.

Daimler Annual Report. (2010). https://www.daimler.com/documents/investors/berichte/geschaeftsberichte/daimler/daimler-ir-annualreport2010.pdf

Daimler. (2011). *Sustainability report 2011*. Stuttgart: Daimler AG. http://sustainability.daimler.com

Daimler. (2012). *Sustainability report 2012*. Stuttgart: Daimler AG. http://sustainability.daimler.com

Diamond, J. (2005). *Collapse. How societies choose to fail or survive*. London: Penguin Books.

Dresner, S. (2008). *The principles of sustainability*. London: Earthscan.

Duwebi, N. (2017) Sustainability integration into project management: Evidence for top SAM listed companies. PhD Thesis, School of the Built Environment and Engineering, Leeds Beckett University

Eccles, R.G., Ioannou, I., Seafeim, G. (2011). The impact of corporate sustainability on organizational process and performance, Working Paper Summaries, Harvard Business School 14/11/11 http://hbswk.hbs.edu/item/the-impact-of-corporate-sustainability-on-organizational-process-and-performance

Ehrlich, P. (1968). *The population bomb. Population control or race to oblivion?* New York: Ballantine.

Elkington, J. (1997). *Cannibals with Forks. The triple bottom line of 21st century business*. Oxford: Capstone Publishing.

Elkington, J. (1999). *Cannibals with Forks. The triple bottom line of 21st century business*. Oxford: Capstone Publishing.

Epstein, M. (2008). *Making sustainability work: Best practice in managing and measuring corporate social, environmental and economic impacts*. Sheffield: Greenleaf Publishing Ltd..

Epstein, M. J., & Roy, M.-J. (2003). Making the business case for sustainability. Linking social and environmental actions to financial performance. *Journal of Corporate Citizenship, 9*, 79–96.

Fiat S.P.A. Sustainability Report. (2011). *Economic, environmental and social responsibility*. Sustainability report 2011. Fiat S.P.A.

Hamani, K., Al-Hajj, A. (2015). A conceptual framework towards evaluating construction contractors for sustainability, the construction, building and real estate research conference of the Royal Institution of Chartered Surveyors RICS COBRA AUBEA, 8–10 July 1015

International Labour Office (2011). A skilled workforce for strong, sustainable and balanced growth: A G20 training strategy, Geneva November 2010, Switzerland. Available at: https://www.oecd.org/g20/summits/toronto/G20-Skills-Strategy.pdf

IUCN, World Conservation Union (1993). *Guide to preparing and implementing national sustainable development strategies and other multi-sectoral environment and development strategies*.

International Institute for Environment and Development, Environmental Planning Group, London.

IUCN, WWF, & UNEP. (1980). *The world conservation strategy*. Switzerland: Gland.

Lamprinidi, S., Ringland, L. (2008). A snapshot of sustainability reporting in the construction and Rea estate sectors. Global Reporting Initiative, GRI Research & Development Series. Available https://www.globalreporting.org/resourcelibrary/A-Snapshot-of-sustainability-reporting-in-the-Construction-Real-Estate-Sector.pdf

Malthus, T. (1798). *An essay on the principles of population*. Oxford: Oxford University Press. (1993).

Mansouri, N. (2016). A case study of Volkswagen unethical practice in diesel emission test. *International Journal of Science and Engineering Applications, 5*(4), 213–216. Available http://www.ijsea.com/archive/volume5/issue4/ijsea05041004.pdf

Merica, D. (2017). Trump dramatically changes US approach to climate change. CNN Politics. March 29, 2017 http://edition.cnn.com/2017/03/27/politics/trump-climate-change-executive-order/

Moorhouse, G. (2005). *Great Harry's Navy. How Henry VIII gave England Sea Power*. London: Weidenfeld & Nicholson.

Niels, F., Jorna, R., & Van Engelen, J. (2005). *The sustainability of "Sustainability" – A study into the conceptual foundation of the notion of "Sustainability"*. Journal of Environmental Assessment Policy and Management, 7(1), 1–33. Imperial College Press.

Oke, A. E., & Aigbavboa, C. O. (2017). *Sustainable value management for construction projects*. Switzerland: Springer.

Organization for Economic Cooperation and Development, OECD (1990). *ISSUESPAPERS: On integrating environment and economics,* Paris.

O'Riordan, T. (1988). *Sustainable environmental management: Principles and practice*. London: Belhaven Press.

O'Riordan, T., & Yaeger, J. (1994). Global environmental change and sustainable development. In *Global change and sustainable development in Europe*. Wuppertal Institute, Nordrhein-Westfalen.

Osborne, G. (2015). *Fixing the foundations: Creating a more prosperous nation*. HM Treasury, CM 9098 July 2015. https://www.gov.uk/government/uploads/system/uploads/attachment_data/file/443898/Productivity_Plan_web.pdf

Peansupap, V., Walker, D. H. T. (2004). Strategic adoption of information and communication technology (ICT): Case studies of construction contractors. In F. Khosrowshahi (Ed.), *20th Annual ARCOM Conference, 1–3 September 2004, Heriot Watt University* (Vol. 2, pp. 1235–1245). Association of Researchers in Construction Management

Pezzey, J. (1992a). *Sustainable development concepts (Rep. No. 11425)*. Washington, DC. http://documents.worldbank.org/curated/en/237241468766168949/pdf/multi-page.pdf

Pezzey, J. (1992b). Sustainability: An interdisciplinary guide. *Environmental Value, 1*(4), 321–362.

Puri, D., & Tiwari, S. (2014). Evaluation the criteria for contractors' selection and bid evaluation. *International Journal of Engineering Science Invention, 3*(7), 2319–6734.

Rees, M. (2003). *Our final century. Will the human race survive the twenty-first century?* London: William Heinemann.

Rockstrom, J., Steffen, W., Noone, K., Persson, A., et al. (2009). Planetary boundaries: Exploring the safe operating space for humanity. *Ecology and Society, 14*(2), 32.

Shipman, T. (2017) 'Less climate concern' key to Brexit trade, Sunday Times, 9 Apr 2017

Silva, G., & De Serio, L. C. (2016). The sixth wave of innovation: Are we ready? *Departamento de Administração, Faculdade de Economia, Administração e Contabilidade da Universidade de São Paulo (USP), 13*(2), 128–135.

Silvius, J., & Schipper, R. (2012). *Taking responsibility: The integration of sustainability and project management*. Retrieved from PM World Today on June 19.

United Nations. (2000). *Cleaning up our mining act: A North-South dialogue*. New York: United Nations.

United Nations Statistical Office. (1992). *SNA draft handbook on integrated environmental and economic accounting*. New York: UN Publications.

Upstill-Goddard, J., Glass, J., Dainty, A. R. J., & Nicholson, I. (2015). Analysis of responsible sourcing performance in BES 6001 certificates. *Proceedings of the Institution of Civil Engineers – Engineering Sustainability, 168*(2), 17–81.

Volkswagen AG. (2012a). *Experience driversity*. Annual report 2012. Volkswagen AG. http://www.volkswagenag.com/content/vwcorp/info_center/en/publications/publications.acq.html/archive-on/icrfinancial_publications!annual_reports/index.html

Volkswagen AG. (2012b). *Sustainability report 2012*. Volkswagen AG.

Washington, H. (2015). *Demystifying sustainability*. Oxford: Earthscan, Routledge.

Watson, J. (2017). Will the energy transition be derailed. UKERC. 28 March 2017. http://www.ukerc.ac.uk/news/will-the-energy-transition-be-derailed-.html

World Bank. (1987). *Environment, growth and development*. Washington, DC: Development Committee Pamphlet 14, World Bank.

World Bank. (1992). *World development report, 1992: Development and the environment*. New York: Oxford University Press.

WWF (2017) A letter to the Secretary of State, Clean Growth Plan 27 March 2017. https://www.wwf.org.uk/sites/default/files/2017-03/Clean%20Growth%20Plan%20letter.pdf

Part II
Case Studies

Chapter 7
A Social-Environmental Interface of Sustainable Development: A Case Study of Ghadames, Libya

Ahmad Taki and Jamal Alabid

7.1 Background

A house is the inhabited space that localises our memories in not only time but also surroundings (Coghlan 2010). The efforts to make buildings socially sustainable have been led by organisations such as Oxford Institute for Sustainable Development (OISD) but with difficulties in measuring the aspects of social sustainability (Lehtonen 2004). In vernacular architecture, the urban structure of the city is set according to the spatial location of locals' social structures (Bramley and Dempsey 2006). Glassie (1990, p.9) stated that "All architecture is the embodiment of cultural norms that pre-exist individual buildings. Vernacular traditions are characterised by a tight correlation between the understanding of these norms by designers, builders and users".

Yet, the remaining social image in traditional buildings and towns, therefore, represents the engagement and participation of man within a building, which is more or less based on an established sociocultural order. Glassie (1990, p.20) stated that "All buildings must be designed with specific cultures in mind. What is right for us is not necessarily right for another culture". However, in practice, the social and environmental dimensions have not been integrated into a holistic approach in the design process, especially in highly dense cities (Bay 2011). Many in literature, including Abarkan and Salama (2000), Chojnacki (2003) and Elwefati (2007) stated that these social and environmental dimensions have been neglected in some developing countries due to ignoring and misunderstanding the values of local architecture and community needs. Golubchikov and Badyina (2012) also believed that there is a reciprocal impact between housing development and the environment where both are related to the climate and ecological change, particularly in contemporary constructions.

A. Taki (✉) · J. Alabid
Leicester School of Architecture, De Montfort University, Leicester, UK
e-mail: ahtaki@dmu.ac.uk

© Springer International Publishing AG, part of Springer Nature 2018   117
M. Dastbaz et al. (eds.), *Smart Futures, Challenges of Urbanisation, and Social Sustainability*, https://doi.org/10.1007/978-3-319-74549-7_7

While these studies clearly show the economic benefits of the concept of compact cities, they also highlight the benefits to the city, human comfort, the indoor environment and living conditions. There was a sense of agreement that climate has been the major factor in shaping old settlements and being so often one of the most determinant in forming its buildings (Ben-Hamouche 2008 and Sundarraja et al. 2009). On the other hand, different groups believe that the cultural and local conditions, including materials and building methods, have a greater impact on how cities were built (Cofaigh et al. 1998). However, lessons acquired from previous generations in traditional architecture dictate that existing social aspects besides climate, materials, topography, land cover and economy should be integrated into the design process of any residential scheme. Yet, the debate of which factors contribute most to shaping urban settlements, specifically between climate and culture, has not been resolved. As such, assumptions rely on a number of variables including the context, the observation angle of the phenomenon and the background of the debater.

According to many in literature, vernacular buildings are more climatically receptive to the environment than modern constructions (Howley 2009). However, in most cases, climate has been the main driver for local builders to identify the traditional building methods and techniques to achieve optimum comfort conditions with limited use of resources (Sundarraja et al. 2009). It can be said that traditional architecture may reflect the vernacularity of a place (sociocultural and historical context of locals) but may not be sustainable if dwellers are not satisfied regardless of the use of locally available materials, methods and techniques, natural resources and addressing environmental issues. This chapter discusses the design priorities of desert architecture responsible for the long-standing nature of the local architecture in spite of harsh environmental conditions while retaining social values and cultural norms. Ghadames housing presents a good example of how humans have learnt balance between the need for culture-specific design and coping with extreme climate conditions, both in syncytium with nature in such way to represent sustainable practices.

7.2   Research Methodology

The methodology is qualitative in nature, using methods of field surveys to investigate existing traditional housing design methods and strategies employed to bring coherence between nature and settlers in consistent systems. A number of traditional houses were visited to evaluate the indoor climate (thermal and visual conditions) through temperature records, drawings and observations of design techniques and methods used in traditional housing. In addition, 12 interviews were conducted during the visit to the old town of Ghadames, to explore in depth, the locals' experience and perception of traditional settlements and whether it has successfully met their basic needs. Although the analysis of the case study is descriptive, a dynamic simulation analysis also is used in an attempt to understand the way these houses were designed to naturally maintain acceptable indoor thermal conditions.

Fig. 7.1 Old and new towns of Ghadames

7.3   Ghadames

The case study is a town located in the Sahara Desert built over 400 years ago on an oasis that lies approximately 630 km to the south-west of Tripoli close to the junction between Libya, Tunisia and Algeria with an altitude of 340–370 m above sea-level (Chojnacki 2003). Although the town was inhabited thousands of years ago, the existing fabric of traditional buildings goes back approximately 200 years and remains as it was. The location between the three countries' borders is of importance to connecting the Mediterranean coast with central Africa. After local residents left the old town in 1985, it was added to the World Heritage List by UNESCO (Fig. 7.1).

According to Zifan (2016) Libyan climate is classified as six climatic zones as defined by the Libyan map of Köppen climate classification. Ghadames lies in the hot arid desert climate. It is characterised as having very low precipitation, far less than the potential evapotranspiration, with large temperature swings particularly during summer months, resulting in low humidity rates (Fig. 7.2). The hot dust-bearing wind known as Ghibli has been the most dominant concern for locals and their regional architecture. The old town settlements represent a basic form of wind-breaks erected in an inclined plane on the leeward of Ghibli surrounded by large green belt of palm trees to give some protection against undesirable summer winds.

7.4   Social Integration and the Spatial Planning of Old Settlements

Ghadames settlements have integrated design solutions that highlight the importance of family privacy long before Islamic rules mandated it. The houses are divided into different zones for both males and females with the ground floor allocated as semipublic, the first floor as semiprivate and then mezzanine and roof levels

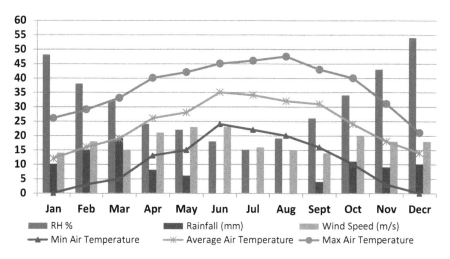

Fig. 7.2 Climate conditions in Ghadames

Fig. 7.3 Social practice and microclimate features in the old town of Ghadames

for family use only. Females so often spend time inside homes and use mainly roofs to move from one house to another (Fig. 7.3). Meanwhile, males spend most of the time outside either taking up commercial or agricultural activities using semi-shaded and shaded alleyways to move around the town. The old town is located in an oasis surrounded by farms with over 36 thousand palm trees (Ealiwa 2000). It can be said that the compact house design was successful in meeting locals' privacy and social needs and allows them to comfortably practise their cultural norms and beliefs.

The lifestyle and city structure necessitated the creation of public squares and baths, and later mosques were added in the vicinity of these town facilities including the market after the Othman invasion. This unique structure illustrated by buildings, streets, alleyways and squares represents the life of its inhabitants and interrelated social relations. It should also be noted that farms surrounding the town dwellings were also hedged by high and thick walls to protect them from being buried by sand. Each of these farms has a separate entrance leading to the main gates of the city, but also squares in between farms exist for meetings and for workers to rest and perform prayers.

In contrast, the new housing was not built according to the social structure and also has no relation to the natural surroundings. Almost each house possesses an outdoor courtyard that is hardly used for any purpose of social life. Although tradition is still strong in Ghadamesian society, the result of isolated dwellings has resulted in rapidly deteriorating social interactions. The connection to public spaces is lost among claims that walking to the mosque several times a day has become difficult. This civilisational and cultural change is taking place not only in Ghadames but across the Arab world and has affected the local communities in almost every aspect of life. Privacy is among the aspects sacrificed in contemporary houses with occupants now modifying their homes to suit their cultural way of life. This particularly resonates with residents of flats.

7.5   Building Form, Layout and Immediate Climate

There is an argument that building forms are required only to respond to the climate, while the climate can influence the building form but does not determine it (Cofaigh et al. 1998). This is in contrast with Konya and Vandenberg (2011) and Ben-Hamouche (2008) who believe that climatic conditions have the greatest influence on building form and design. They give an example of buildings found in the tropics which should differ from those situated in temperate regions. To a large extent, the sun's path determines the form of buildings, and similarly, the traditional compact design in hot regions has explicitly been adopted to provide protection from the harsh outdoor environment. In traditional urban settlements, the choice of building form, structure, construction methods and materials is clearly related to local climate conditions. The courtyard style of housing has been always the example of hot climate dwellings (Nikpour et al. 2012, Leylian et al., 2010, Abarkan and Salama 2000, and Edwards et al. 2006).

Understanding the features and ecological values of the microclimate early in the design process could help reduce energy expenditure and excessive reliance on mechanical systems of heating, cooling and other operational systems (Ralegaonkar and Gupta 2010). Traditional architecture in hot, dry territories is so often inwardly oriented, characterised by a central courtyard (Hyde 2008). The urban fabric of the old town of Ghadames is the epitomes of adaptation to the harsh and extreme climate conditions. The building form is highly compact and tightly interwoven, built on an incline, centralised to the south-west of the oasis, forming a large complex of townhouses (Fig. 7.4).

The form of the house and relatively small plot area ($25-50$ m^2) allows for minimum exposure to solar radiation. The spatial organisation of rooms varies according to the privacy level and functionality. It is, therefore, constructed over three storeys

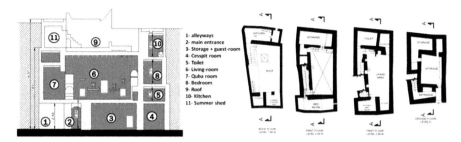

Fig. 7.4 A typical traditional house in Ghadames

to accommodate a fully sheltered ground floor, consisting of the main entrance with stairs leading to the central living hall, guest room, storage and cesspit room. The first floor is a semiprivate family area centralised by a living room surrounded by a number of bedrooms. The central hall is built with steps leading to a mezzanine level consisting of other private bedrooms. Stairs lead up to the roof level where kitchen is found as well as shed space often used in the summer nights. It was noted that almost all houses in the old town are very similar in terms of space arrangement and organisation, and they only differ in size and number of rooms as well as the exposure level.

7.6   Climate and Comfort in Traditional Architecture

Enclosure in buildings can be a key design feature in promoting the microclimate and indoor comfort conditions and consequently have an effect on the heating or cooling loads (Straube 2012). Traditional dwellings are of three storeys intertwined above alleyways allowing natural daylight and ventilation to pass through regularly spaced light wells. The surrounding vegetation is integrated into the land to play a role in moderating the microclimate of the town and protecting it against sandstorms. The double height ceiling of the living room, ranging from 4 to 5 m, acts as moderator between the exposed roof and interiors. The roof aperture found in the living room is responsible for most of the daylight and ventilation brought to the house. According to Mezughi and Adawi (2003), the indoor environment of the old town is relatively cooler during daytime due to less direct solar heat gain and warmer at night because of the high thermal capacity of its building fabric. In addition, warm and cool air enters the house from different places within the passageways owing to the location of the light-well junctions and the fact that different pressure zones are created in street level due to variation in exposure to direct sun causing air movement driven by buoyancy force.

A considerable portion of solar radiation is reflected by the white external surface, and the remaining is absorbed by the relatively thick and high thermal mass materials. Shading methods also ensure less exposure to direct sun. Dwellings differ

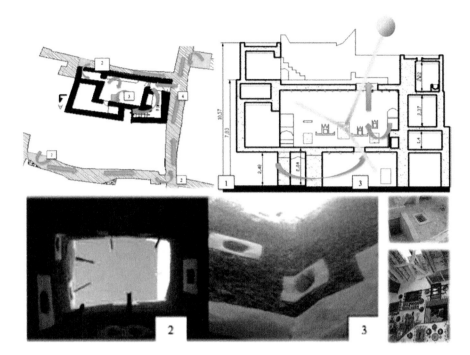

Fig. 7.5 Natural ventilation during daytime in traditional settlements of Ghadames

in height and in the organisation of the rooms commonly found above roof level. These elements create and extend the time of shading, while parapet walls also act as shading devices (Fig. 7.5).

The first floor is noticeably larger in area than other floors as it extends to cover the pathway. The upper floor is allocated for the kitchen to ensure heat and smoke generated are driven away from the house, while the shed often faces north. The use of reflective materials such as mirrors helps distribute sufficient daylight deeper into the surrounding rooms. The relatively small openings and low ceiling height of rooms in vernacular architecture were commonly practised especially in hot climates. Cofaigh et al. (1998) stated that the height of a room must be controlled by its length to produce a better repose feeling. They also believed that if small windows are correctly placed in conjunction with white ceilings, daylight is optimally delivered across the room together with the maintenance of thermal comfort applicable throughout the year. Importantly, this concept saves the need for blinds, providing a good level of sunlight in the room while avoiding the heat of summer days.

The current study found that there are slight differences between the outside air temperature in old and new towns of Ghadames which may explain the difference in the microclimates of the two areas. Summer nights in the Sahara Desert are colloquially known as winter of the summer where temperatures sometimes drop to less than 10°C. It is therefore very important to minimise the absorption of warmth

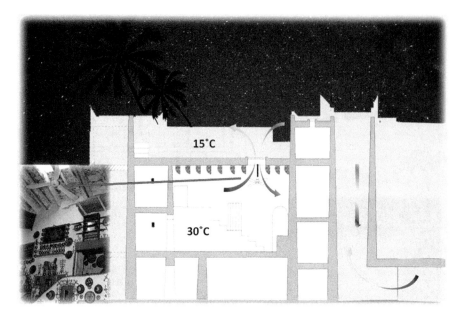

Fig. 7.6 Night ventilation and thermal mass strategies

by fabric during the day through night purge ventilation and the use of fans to enhance air circulation. Cooler night air is heavier in density and therefore sinks down the house via the roof aperture, forcing warming air upwards with the assistance of fans (Fig. 7.6).

Natural light or sunlight is a vital component to life that plays a primordial function in humans' biological and psychological systems (Boubekri 2008). In fact, there are many considerable factors affecting the amount of the daylight available, including geographical location, weather conditions, time of the day and year and the building design form, layout and window design (size, position and orientation) (Alabid et al. 2016). The design principles of daylight should reflect the amount needed for the space as well as the expected heat generated. In addition, human well-being and productivity should be considered as a core concern of daylight principles and designers.

In the case of traditional dwellings in Ghadames, sunlight's light is often undesirable because of the heat, and hence direct sunlight is highly minimised, relying on diffused light instead. The roof skylight brings not only fresh air and daylight but also proximity to the solar cycle. Reflective surfaces compensate for the relatively small roof opening dispersing light into surrounding rooms. Windows integrated into roofs in this way ensure the daylight can enter the house with minimum exposure to outside. This also makes it possible for dwellings to stand wall to wall, further minimising sun exposure and maximising house size (Fig. 7.7).

At street level daylight is distributed in a way that maintains good visual connection to indoor and outdoor environment without compromising pedestrian thermal

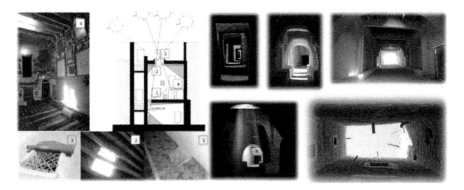

Fig. 7.7 Daylighting strategies in the old settlements of Ghadames

comfort. Semi-shaded alleyways mean that the sky is visible through light wells while protected pedestrians from its heat. The white finishing paint on the roofs and streets also contribute to increasing brightness of the space.

The water source, known as Eyin Alfaras, is a natural spring supplying the whole town with potable water through a number of streamline systems of canals that pass underneath the urban structure (Fig. 7.8). This water system plays a great part in the stability and creation of the microclimate in the old town of Ghadames. Water adds moisture to the otherwise dry climate and contributes to climate cooling in covered streets. Moreover, water flows by a gravity-fed system stimulating air current to circulate indoors, cooling and humidifying the indoor climate by latent heat exchange which in turn affects occupant thermal comfort.

7.7   Social-Environmental Equilibrium in Built Environment

Understanding local context and conditions (geographical location, history, economy, climate, topography and general environmental conditions) is key for appropriate design related to architectural form, space, structure and selection of materials and even creating sustainable communities. As a matter of fact, humans have developed physiological and behavioural attributes over millennia as a result of exposure to varied climatic regimes although it is subjected to culture, place or time. On both urban and building design scales, Ghadames traditional settlements showed to be highly climatically responsive and also fulfilled and respected the social values of local society at least for the time at which they were constructed. Old settlements provided considerable passive cooling strategies, good use of daylight and moreover a place where inhabitants are able to work and rest throughout the day, protected from the scorching sun and sandy hot winds. Sassi (2006) mentions strategies for sustainable architecture highlighting the significance of living in harmony with nature as it plays a fundamental part in improving quality of life.

Fig. 7.8 Natural spring water and microclimate in the old town of Ghadames

It is, therefore, clear that the builders of the old settlements thoroughly both the indoor and outdoor environments to be able to meet more than merely the basic needs. Indubitably, climate is one of the main driving and dominant forces in the solutions used to build old settlements in this oasis. However, there are other innate factors that must contribute to generate the forms of traditional architecture of the town. These consist of sociocultural and economic aspects, religion, constructive techniques, materials and resources available. The town indeed expresses its identity urbanistically, architecturally and behaviourally and is an ideal example of man living in harmony with nature (Fig. 7.9).

7.8   Indoor Environment and Occupants' General Perception

During the visit to the three case studies in the old town, interviews were conducted with a number of professionals. Furthermore, temperature was recorded and participants' thermal sensation was assessed. Temperature measurements were taken in the central hall on the same day as the interviews, every 15 minutes for an hour as demonstrated in Table 7.1. The Table readings are an average of all records over that hour. Dear et al. (2013) found that comfort can be achieved at temperatures between 25.5 °C and 29 °C with air velocity between 0.2 and 1.4 m/s. However, Arens et al. (2013) found that the majority of participants were comfortable at air speed ranging

Fig. 7.9 Culture and environment balance in traditional settlements of Ghadames

Table 7.1 Indoor temperature readings in traditional house

Physical Parameter	Case 1	Case 2	Case 3
Local time	14:00 pm	15:10 pm	16:05 pm
Globe temperature	30.0°C	33°C	31.0°C
Wet bulb temperature	20.0°C	21.5°C	19.8°C
Air speed	0.11 m/s	0.04 m/s	0.07 m/s
Air temperature	31°C	32.2°C	31.5°C
Relative humidity %	36%	38.7%	33.1%

from 0.05 to 1.8 m/s, relative humidity of 60%–80% and temperature between 26°C and 30°C.

Surface temperatures of walls, floors and roofs were measured to find out the mean radiant temperature (MRT) that most likely denotes how thermally comfort the inhabitant is. Table 7.2 shows surface temperature records inside the living room during the visit to the *Dar Amazagra* house in the old town of Ghadames.

In some studies such as Ealiwa (2000) and Gabril (2014), comfort temperature was found to be highly dependent on outdoor temperature and also on surface temperatures. An equation has been derived to determine the comfort temperature based on MRT for Ghadames terrain as following:

$$T_{comf} = 0.46T_{mrt} + 16.7 \pm 2$$

Table 7.2 Comfort temperature inside traditional dwellings of Ghadames

Surface	$T_{wall\ 1}$	$T_{wall\ 2}$	$T_{wall\ 3}$	$T_{wall\ 4}$	T_{Roof}	T_{Floor}	MRT
Temperature	31.9°C	31°C	29.4°C	32.3°C	31.2°C	28.6°C	*31.7°C*

Table 7.3 Professionals' views on traditional dwellings

Opinion	Reason	No. of interviewees	%
Agree	Reduces the building cost and land use Suitable for climate Similar concept has been already tested	10	83.33
Disagree	Unless considering the new space layout More likely to be unaccepted in the society	2	16.66

According to the MRT found in traditional houses, the comfort temperature can be estimated as

$$T_{comf} = 0.46 \times 31.7 + 16.7 \pm 2 = 31.28°C \pm 2$$

This result is in agreement with a previous study carried out in 2013 to investigate thermal comfort boundaries by surveying a number of traditional houses in Ghadames. Occupants were found to be most thermally comfortable at temperatures of 29°C–32°C (Alabid et al. (2014)).

During the interview with a number of householders, the possibility of reusing of the traditional house concept in future housing designs was discussed while illustrating how it could contribute to reducing the constructional and operational costs. 83% agreed to this and were convinced of its robustness as shown in Table 7.3. Only 17% disagreed with the reuse of the concept by the reasoning that the interior design and layout of the traditional house concept would have to be entirely changed to meet todays' requirements. Furthermore, they highlighted the fact that the younger generation are less likely to be able to adapt to this traditional way of life. The table shows interviewees' opinions and reasons for agreeing to the statements.

7.9 Conclusion

In vernacular architecture, culture and climate are no doubt the dominant concerns of building professionals. Together with other factors including availability of construction materials, resources and even the technical capacity, they have influenced the final outcomes of the local settlements. Builders of Ghadames old town are successfully dealing with one of the harshest climates of the Sahara Desert bringing harmony between natural and man-made systems while still representing local architecture and lifestyle. This study carried out field surveys to investigate how effective the old settlements of Ghadames are in addressing environmental and social aspects and, equally, the perspective of the locals regarding the current indoor and outdoor conditions. Interviews with locals highlighted the effectiveness of the old settlements with particular respect to social life and family privacy reflected in both the indoor and outdoor climates. Also, it should be noted that indoor thermal comfort was achieved, the majority of the time owing to the minimal exposure to sunlight both indoors and at street level. Furthermore, the combination of natural ventilation and thermal mass in the traditional homes together with night purge strategies helps maintain comfortable indoor temperatures. Water plays a significant role in moderating microclimate thermal conditions and creates a synergic system promoting community cohesion and outdoor interaction. Finally, the majority of building professions agree on the concept of reviving the old settlement design strategies and adapting them to modern life as a sustainable architectural method without compromising cultural values and traditions.

References

Abarkan, A., & Salama, A. (2000). Courtyard housing in Northern Africa: changing paradigms. Paper presented at the ENHR 2000 Conference, Housing in the 21st century: Fragmentation and Reorientation, 26–30 June 2000, Gavle, Sweden.

Alabid, J., Taki, A., & Painter, B. (2016). Control of daylight and natural ventilation in traditional architecture of Ghadames, Libya. In *21st Century human habitat: issues, sustainability and development*. Akure: Federal University of Technology Akure.

Alabid, J. M., Taki, A., & Cowd, B. (2014). Desert architecture review of Ghadames housing in Libya. In *First International conference on energy and indoor environment for hot climates* (pp. 240–247). Doha: ASHRAE.

Arens, E., et al. (2013). *Final report air movement as an energy efficient means toward occupant comfort*. Berkeley: California.

Bay, P. J.-H. (2011). Social and environmental dimensions in ecologically sustainable design: towards a methodology of ranking levels of social interactions in semi- and open spaces in dense residential environments in Singapore, pp. 162–177.

Ben-Hamouche, M. (2008). Climate, cities and sustainability in the Arabian region: Compactness as a new paradigm in Urban Design and planning. *Arch. Net-IJAR: International Journal of Architectural Research, 2*(2), 196–208.

Boubekri, M. (2008). *Daylighting, architecture and health first edit*. Oxford, UK: Elsevier Ltd..

Bramley, G. & Dempsey, N. (2006). What is 'social sustainability', and how do our existing urban forms perform in nurturing it. … http://www.City-form.Com…, (April), pp. 1–40.

Chojnacki, M. (2003). Traditional and modern housing architecture and their effect on the built environment in North Africa. In *Methodology of housing research* (pp. 1–22). Stockholm: Royal Institute of Technology (KTH).

Cofaigh, E. O., Olley, J. A., & Lewis, O. (1998). *The climatic dwelling: An introduction to climate-responsive residential architecture*. London, UK: James & James.

Coghlan, N. (2010). New architectures of social engagement. Aesthetica magazine, (October), pp. 22–25.

Dear, R., et al. (2013). Progress in thermal comfort research over the last twenty years. *Indoor Air, 23*, 442–461.

Ealiwa, A. (2000). *Designing for thermal comfort in naturally ventilated and air conditioned buildings in summer season of Ghadames, Libya*. Leicester: De Montfort University.

Edwards, B., et al. (2006). *Courtyard housing; past, present & future 1st edition*. Abingdon, UK: Taylor & Francis e-Library.

Elwefati, N. (2007). *Bio-climatic Architecture in Libya: Case studies from three climatic regions*. MSc thesis, Architecture Department, Middle East Technical University, Ankara, Turkey.

Gabril, N. (2014). *Thermal Comfort and Building Design Strategies for Low Energy Houses in Libya: Lessons from the vernacular architecture*. PhD thesis, University of Westminster, London, UK.

Glassie, H. (1990). Architects, vernacular traditions, and society. TDSR, 1. Available at: http://iaste.berkeley.edu/pdfs/01.2b-Spr90glassie-sml.pdf.

Golubchikov, O., & Badyina, A. (2012). In M. French (Ed.), *Sustainable housing for sustainable cities*. Nairobi: UN-Habitat.

Howley, P. (2009). Attitudes towards compact city living: Towards a greater understanding of residential behaviour. *Land Use Policy, 26*(3), 792–798.

Hyde, R. (2008). *Bioclimatic housing: Innovative designs for warm climates*. London: Cromwell Press.

Konya, A., & Vandenberg, M. (2011). *Design primer for hot climates*. Reading, UK: Archimedia Press Limited.

Lehtonen, M. (2004). The environmental–social interface of sustainable development: Capabilities, social capital, institutions. *Ecological Economics, 49*(2), 199–214.

Leylian, M., Amirkhani, A., Bemanian, M., & Abedi, M. (2010). Design principles in the hot and arid climate of Iran, the case study of Kashan. *International Journal of Academic Research, 2*(5).

Mezughi, M., & Adawi, M. (2003). *Consultancy report submitted to UNDP for rehabilitation of the old town*. Ghadames.

Nikpour, M., et al. (2012). Creating sustainability in central courtyard houses in desert regions of Iran. *International Journal of Energy and Environment, 6*(2), 226–233.

Ralegaonkar, R. V., & Gupta, R. (2010). Review of intelligent building construction: A passive solar architecture approach. *Renewable and Sustainable Energy Reviews, 14*(8), 2238–2242.

Sassi, P. (2006). *Strategies for sustainable architecture*. New York: Taylor & Francis e-Library.

Straube, J. (2012). Insight the function of Form: Building shape and energy. *Building Science, BSD-061*, 1–4.

Sundarraja, M.C., Radhakrishnan, S. & Priya, R.S. (2009). Understanding vernacular architecture as a tool for sustainable built environment. 10th National Conference on Technological Trends (NCTT09), pp. 249–255.

Zifan, A. (2016). Libya map of Köppen climate classification, Wikimedia Commons.

Chapter 8
The Undervaluation, but Extreme Importance, of Social Sustainability in South Africa

Elizelle Juanee Cilliers

8.1 The Doctrine of Sustainability and Spatial Planning

Sustainable development is contextualised as the trade-off among social, economic and ecological objectives of conservation and changes (Goel and Sivam 2014:61). Sustainability as a universal ambition recently became a land use issue, encapsulated in the United Nations Sustainable Development Goals number 11, calling for inclusive, safe, resilient and sustainable cities and human settlements (United Nations 2017). This is no easy task at hand, in light of increasing urbanisation and development pressure, poverty and the growing importance of the green hype. Land use planning is therefore set as an arena in which conceptions of sustainable development are contested (Godschalk 2004:6), considering systems thinking (Richmond 1993) and ever seeking to balance the three interrelated dimensions of environmental sustainability, economic sustainability and social sustainability.

The doctrine of sustainable development derived from the economic discipline, captured in research of political economist Thomas Malthus in the early 1800s (Basiago 1999:145). Economists focused on the efficiency of resource usage and ignored the dilemma of resources depletion, until 1972 when the first influential work was published that questioned the sustainability of the paradigm of world economic development (Basiago 1999:146). The apprehension that industrial production is eroding natural resources upon which economic development depend led to the UN Conference on Human Environment in Stockholm in 1972, bringing representatives of developed and developing countries together for the first time in history, to debate humanity's right to a healthy and productive environment (Basiago 1999:146). The term 'sustainable development' was coined in the World Conservation Strategy drafted by the United Nations Environment Programme

E. J. Cilliers (✉)
Urban and Regional Planning, Unit for Environmental Sciences and Management, North-West University, Potchefstroom, South Africa
e-mail: Juanee.cilliers@nwu.ac.za

© Springer International Publishing AG, part of Springer Nature 2018
M. Dastbaz et al. (eds.), *Smart Futures, Challenges of Urbanisation, and Social Sustainability*, https://doi.org/10.1007/978-3-319-74549-7_8

(UNEP) and the International Union for the Conservation of Nature (IUCN) in 1980, set to be advanced through conservation.

It was in 1987 that the United Nations World Commission on Environment and Development, chaired by Gro Harlem Brundtland of Norway, recalled the concept of sustainable development and defined it as 'development that meets the needs of the present without compromising the ability of future generations to meet their own needs' (WCED 1987). Thereafter, various worldwide events enforced the principles of sustainable development, leading to the most recent Sustainable Development Goals. The advent of sustainability in science has since led planners to apply evolving notions of sustainability to the debate on successful city planning (Basiago 1999:148).

As such, concepts of environmental sustainability and social sustainability are now considered equally important to the original economic sustainability objective raised by the Political Economics in the 1800s. Environmental sustainability is enforced by environmental considerations which have become an integral part of developmental thinking and decision-making based on the expanded scientific understanding that environmental and related ecological systems are crucial to achieve urban sustainability (Thomas and Littlewood 2010:212; Wright 2011:1008). As such, ecological principles became a sine qua non for effective designs and solutions for cities (Forman 2013), providing forward-thinking solutions to spatial planning (Landscape Institute 2013:1). Social sustainability, on the other hand, remains an elusive concept, often encapsulated in the concepts of social cohesion and social capital (Dixon and Woodcraft 2013), as two interrelated ideas (Carrasco and Bilal 2016:127). However, urban theory provides no consensus as to which human settlements embody 'sustainability' (Basiago 1999:148) and possibly the reason why social sustainability is renowned in literature as the least understood of the three dimensions of sustainability.

8.2   A Spatial Perspective on Social Sustainability

'Development, conservation and planning all exist in combination as part of a public corporatist agenda for pre-figuring the general good of communities within society, and for the benefit of individuals' (Riddel 2004:49). Planning interventions that enforce 'natural' change patterns are justified on the presumption that such intervention will make a useful difference to people. The core business of the profession of planning is engraved in social sustainability. However, planning traditionally focused on economic growth and more recently on environmental conservation, and the notion of social sustainability was included as a mere spin-off. This is evident from the classification of sustainability in terms of ecology, equity and economy (refer to Fig. 8.1) where identified tensions (development conflict, resource conflict and property conflict) primarily related to the conflict between pro-developmental approaches versus pro-environmental approaches as explained by Cilliers (2009).

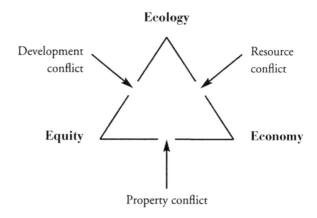

Fig. 8.1 Conflicts among sustainable development values (Source: Godschalk (2004:6))

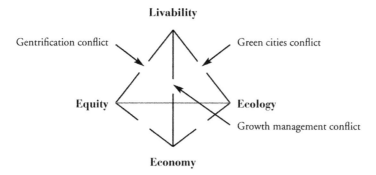

Fig. 8.2 The sustainability-liveability prism and conflicts (Source: Godschalk (2004:9))

The social dimension is underplayed in this regard. Research conducted by Godschalk (2004:8) called for the inclusion of 'liveability' as part of this sustainability representation, creating a three-dimensional figure including the four values of ecology, equity, economy and liveability (refer to Fig. 8.2). Research acknowledged that such social perspective will initiate new conflicts in terms of growth management, green cities and gentrification but would reclaim the social sustainability focus.

From a spatial planning perspective, the connections between social sustainability and the opportunities provided by the physical environment are becoming more apparent as land use management are set to guide urban growth to provide high-quality living environments. However for such to realise, planners ought to comprehend the fine balance needed between natural and built environments and employ context-based planning, founded on community needs. This chapter argues that context-based planning forms the core point of departure for realising sustainable development, as context-based planning responds to social sustainability objectives, taking needs of specific communities and cultures into account and tailoring the planning of their environment, to reflect such.

Context-based planning within the broader social sustainability umbrella is an especially pertinent field of research in post-apartheid South Africa in recognition of the challenges and inequalities faced by South African society, aptly exhibited in urban landscapes where millions reside in shantytowns and informal backyard rental accommodation (Lategan 2016). Ironically, social aspects are not given the same importance as environmental and economic aspects (Goel and Sivam 2014:1). Accordingly this chapter considered sustainability from a South African perspective.

8.3   Sustainability from a South African Perspective

The South African perspective on sustainability was drawn from a qualitative enquiry into a literature review, followed by a reflection on six individual studies that were conducted in South Africa between 2014 and 2017 on diverse planning-related themes. The literature review scrutinised the diverse benefits of green spaces in cities, relating to the three dimensions of sustainable development. It was illustrated that although internationally accepted, many of these studies have not been conducted within a South African context, and the validity thereof was questioned. In other cases contradictions with theory were eminent, and such was elaborated on in Sect. 3.2 as empirical investigation into the specific case studies. The reflection of the six individual cases provided insight into the unique social context and the impact that such have on planning studies. None of these cases aimed to investigate social issues, but findings illustrated deviations from theory, initiated by the unique social context.

8.3.1   Literature Review Concerning three Dimensions of Sustainability

A literature review conducted for a research project funded by the South African Cities Network in 2016 (Cilliers and Cilliers 2016) employed theory-based sampling as part of qualitative enquiring into the quantification of green space values. It identified various studies that captured the value of green spaces in terms of social benefits, environmental benefits and economic benefits and classified such for both household- and neighbourhood scales. The investigation identified that most of these studies were conducted internationally, confirming the limited data available to quantify the green values within local South African context. Table 8.1 captures the individual studies that were evaluated as part of the literature review in terms of the three dimensions of sustainability, where local South African studies are illustrated in bold text.

Table 8.1 Literature review of research conducted on the valuation of green spaces.

Benefits		Household level	Neighbourhood level
Direct economic benefits	Financial	Higher property prices internationally (Perman et al. 2003; Luttik 2000)	Enhanced competitiveness of places (Stigsdotter 2008)
		Raised property prices locally (Schäffler et al. 2013:126, Irwin 2002)	Market values (Van Leeuwen et al. 2009)
		Higher neighbourhood values locally (Cilliers et al. 2013)	Lower storm water costs (Stiles 2006)
		Positive impact on production (Cilliers et al. 2012)	Lower emissions (Bolund and Hunhammar 1999).
		Increase in economic well-being (Beck 2009:240; Chen and Jim 2010)	Higher marketability of areas (Woolley et al. 2003)
		Lower maintenance costs (Stiles 2006)	Increased tourism spending (Swanwick et al. 2003)
		Contribute to house-buyers preferences (Bolitzer and Netusil 2000).	Reduction in costs of pollution control (Sutton 2006)
		Higher property values (Crompton 2001; Fausold and Lilieholm 1999)	More inward investment (CABE Space 2005)
		Valuation of grasslands (De Wit and Blignaut 2006)	Favourable image of the place (Woolley et al. 2003)
			Boost retail sales (Woolley et al. 2003)
			Improving the legibility of the city or neighbourhood (Stiles 2006)
			Multidimensional & policy value (Van Leeuwen et al. 2009)
			Open space values (Turpie and Joubert 2001; Roberts et al. 2005)
			Lower costs of artificial wetlands (Hardin 2001)
Indirect economic benefits	Social	Enhance community cohesion (Ulrich and Addoms 1991)	Enhance urban renewal (Sutton 2006)
		Better quality living space (Cilliers et al. 2010)	More social capital (Cilliers et al. 2010)
		Aesthetic enjoyment (Kong et al. 2007)	Aesthetic values, visual amenities (Natural Economy NW 2007)
		Recreation opportunities (Tyrvainen 1997)	Cultural values, cultural amenities (Natural Economy NW 2007)
		Leisure possibilities (Ulrich and Addoms 1991)	Identity of space (Luttik 2000)

(continued)

Table 8.1 (continued)

Benefits		Household level	Neighbourhood level
		Health benefits (Luttik 2000, Van den Berg et al. 2007)	Better neighbourhood relationships (Roger 2003)
		Contribute to well-being (Cities Alliance 2007; Luttik 2000)	Greater substitution value (Van Leeuwen et al. 2009)
		Positive perception (Kazmierczak and James 2008)	Enhance urban liveability (Caspersen et al. 2006)
		Psychological restoration (Kuo 2003, Van den Berg et al. 2007;)	Crucial to children's social & cognitive development (Sutton 2006)
		Stress relief (Hansmann et al. 2007; Herzele and Wiedemann 2002).	Establishing a sense of place (Stiles 2006)
		Positive social impact on children (Taylor et al. 2002)	
		Facilitation social contact and communication (Stiles 2006)	
		Positive assimilation of values and moral attitudes (Sutton 2006)	
		Access to experience (Kazmierczak and James 2008)	
		Recreational value (Turpie and Joubert 2001)	
	Environmental	Access to clean air and (Kong et al. 2007)	Higher biodiversity (Cilliers et al. 2013)
		Noise reduction (Bolund and Hunhammar 1999, Stiles 2006)	Ecological functions and ecosystem services (Stiles 2006)
		Enhance natural settings for play (child-friendly spaces)	Sustainable environments (Bedimo-Rung et al. 2005;)
		Increase intrinsic natural value (Van Leeuwen et al. 2009)	Habitat protection and provision (Cilliers et al. 2012, Stiles 2006)
		Life-support value (Van Leeuwen et al. 2009)	Lower air pollution levels (Natural Economy North West 2007)
		Air and water purification (Sutton 2006)	Water management (Natural Economy North West 2007)
		Quiet environments (Tyrvainen 1997)	Storm water management (Stiles 2006)
		Clean water and air (European Union 2013:6)	Improved land quality (European Union 2013:6)
		Rainwater retention (European Union 2013:6)	Lower carbon dioxide (McPherson et al. 2002)
			Climatic amelioration (Stiles 2006:11)
			Sustainability (Hodgkison and Hero 2002)
			Carbon sequestration (Schäffler and Swilling 2013)

In terms of environmental sustainability, several of these international research and theories could be translated to a local South African context, as proven to be true, despite South Africa being a developing country. Such include the enhanced biodiversity provided by green spaces in cities (Cilliers et al. 2013), the myriad of ecosystem services that these spaces offer to human societies (Stiles 2006:30), the provision of habitats for wild plants and animals (Stiles 2006:31) and increased intrinsic natural values (Van Leeuwen et al. 2009). There were, however, various research and theories identified that could not be verified for the South African context such as enhanced water management approaches provided by green spaces (Sutton 2006), a concept that has not received much attention in South Africa, as well as the calculation of street tree costs (tree planting, irrigation and other maintenance) versus calculated benefits (energy savings, reduced atmospheric carbon dioxide, improved air quality and reduced storm water runoff), to estimate net benefits of green spaces (McPherson et al. 2002). Such studies have not been conducted comprehensively in the African context, and the validity thereof could thus not be confirmed.

A similar trend was evident in terms of economic sustainability. The economic benefits of green spaces often refers to market values (Van Leeuwen et al. 2009) determined through hedonic pricing methods that place the attention on the impact of land use on surrounding properties (Irwin 2002) and the valuation of property prices linked to attractive environmental spaces, in comparison to properties in less favourably located areas (Luttik 2000:1). Research that could be related to the South African context included evidence on enhanced tourism activities (Swanwick et al. 2003) and increased inward investment in the area (CABE Space 2005), especially in tourist destinations. No local studies have been conducted to validate international findings relating to lower traditional infrastructure costs as a result of green space provision (Stiles 2006), lower emissions (Bolund and Hunhammar 1999) and increased economic well-being (Beck 2009:240). The most evident contradiction to international acclaimed theory was that of increased property values, due to proximity to green spaces (Cilliers et al. 2012; Woolley et al. 2003; Van Leeuwen et al. 2009). This case was further explored as part of the empirical investigation in Sect. 3.2.

In terms of social sustainability, an even greater disparity was evident between the international and South African context. Although access to and experience of nature are accepted as issues influencing human physical and psychological health and well-being (Stiles 2006:32), comprehensive studies have not been conducted in a South African context, especially considering the unique characteristics of urban and rural areas. The positive perception of urban green space values (Kazmierczak and James 2008; Kuo 2003; Chiesura 2004) are well described in international literature, linked to positive assimilation of values and moral attitudes (Sutton 2006) and enhanced urban liveability (Caspersen et al. 2006). However, the issues of safety and social status seemed to place an entirely different dimension to the South African perspective on social sustainability, not acknowledge in the international planning arena. Such is elaborated accordingly in Sect. 3.2.

8.3.2   Reflection on South African Case Studies and Unique Social Considerations

The reflection of the six individual cases conducted between 2014 and 2017 on diverse planning-related themes in South Africa provided insight into the unique social context and the impact that such have on planning studies. None of these cases aimed to investigate social issues, but findings illustrated deviations from theory, initiated by the unique social context. It confirmed that urban planning is unavoidably context defined, and planning ideas cannot be based on general applicability. The reflection on these case studies were thematically classified, relating to (3.2.1) the adequate knowledge and contextualisation of concepts (3.2.2), contradicting results of the proximity principle in urban areas of South Africa (3.2.3), contradicting results of compensation hypothesis in rural areas of South Africa and (3.2.4) other social issues that impart sustainability.

8.3.2.1   Knowledge and Contextualisation of Concepts

Misunderstanding and *misinterpretation* of concepts are often reasons for exclusion in practice. The concept of green spaces could serve as a worthy example to illustrate this challenge in terms of the South African context. The concept of 'green spaces' is well defined in the international literature as a qualitative open space with specific functions connected thereto (McConnachie and Shackleton 2010). However, in South African planning literature, official policies and databases, the concept of 'green spaces' is often referred to as 'open space' by including definitions of 'developed and undeveloped green space' (Schäffler et al. 2013:3). Open space could range from a vacant site to neglected natural area to sports field.

A quantitative community survey conducted in 2015 (Veiga 2015) in the local Fleurhof area, situated south west of Johannesburg in South Africa, confirmed such statement; 322 questionnaires were completed by residents (3.19% sample size) of the said area to capture their knowledge and perspective regarding green spaces in the area. Data of the convenience sample were statistically interpreted and p-values reported for completeness sake. Chi-square tests and symmetric measures illustrated statistical significant association between different questions, where $p < 0.005$. Findings revealed that 93% of participants considered environmental issues to be important, but only 39% of participants showed a good understanding and knowledge of the concept of green spaces. Statistical analysis identified that younger generations were more informed about green spaces with an effect size of 0.248.

Another qualitative study conducted in 2016 (Huston 2016) investigated the understanding of professional planners relating to green spaces and green infrastructure. Based on the same statistical method as the 2015 study, 13 questionnaires were completed by purposefully selected participants. 69% of participants illustrated knowledge relating to the concept of green spaces, but upon qualitative investigation, only 23% proofed to substantiate such understanding of the concept. Of the respondents working in the private sector, 40% claimed to never have included green infrastructure as part of spatial planning projects. The research illustrated the misunderstanding and misconception of the definition of green spaces and green infrastructure among professional planners, and the unevenly distributed use of green infrastructure in practice. An evident correlation could be drawn between lack of knowledge and realisation of the concept of green spaces (and green infrastructure) in practice.

8.3.2.2 Contradicting Results of Proximity Principle in Urban Areas

As proximity to green space was proven to be a key factor in international residential value (Konijnendijk et al. 2013:21), *a pilot study was conducted in 2016 in Potchefstroom* (Cilliers and Cilliers 2016), situated in the North-West Province of South Africa, to test applicability of the proximity principle in local context. The proximity principle states that open spaces, in general, raise the value of nearby properties (Brander and Koetse 2011; Konijnendijk et al. 2013:21). Five residential areas, inclusive of 188 properties, were purposefully selected based on their proximity to green space and was a refinement of previous research conducted by Cilliers (2010) and Cilliers et al. (2013), as illustrated in Fig. 8.3.

Research sites were not limited to a specific green space but ranged from recreational green spaces to aesthetic green spaces. Only residential properties (zoned residential 1) in the more affluent areas of Potchefstroom were selected as previous studies indicated that the demand and supply for ecosystem services differ between the residential areas along a socio-economic gradient (Cilliers et al. 2013). The function and accessibility to these spaces were considered in terms of safety concerns as a factor impacting on value. Three zones within each of these areas were selected and sampled according to location and distance from the green space. Figure 8.4 illustrates an example of the zones selected in each of the areas.

The residential property prices were based on the municipal property valuations (Tlokwe City Council Valuation Roll) for the period 2009/2013, as provided by the local municipality (Tlokwe City Council 2010). The price per square meter of each property was determined and compared, and a mean value was determined for each zone. Data were analysed in terms of (1) ANOVA effect sizes, (2) ANOVA p-values, (3) ANOVA between means p-value and (4) Kruskal-Wallis p-value. The null hypothesis assumed that all areas should have the same property value irrespective of their distances from the green space. Significant differences would reject the hypothesis, and in such cases, Unequal N Honestly Significant Difference (HSD) test was used to compare the sample means pair wise with that of every other sample. Comparisons between Zone 1 and Zone 2, as well as between Zone 1 and Zone 3, in terms of the ANOVA effect sizes illustrated a large practical significant difference (≈ 0.8) between the mean, as well as the effect size within four of the five areas. Three of the five areas indicated that there is a statistically significant difference between the means ($p < 0.05$ ANOVA analysis) and between the groups ($p < 0.05$ Krusal-Wallis analysis). In all five areas, Zone 1 had a lower price per square meter than in comparison to both Zone 2 and Zone 3 (Cilliers and Cilliers 2016).

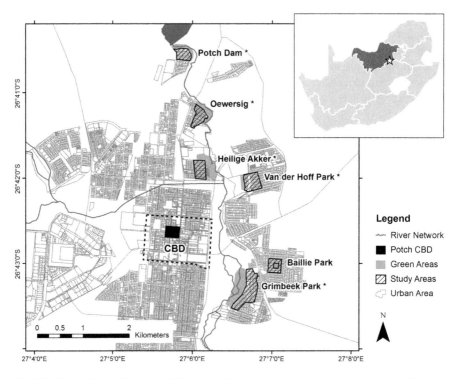

Fig. 8.3 Greater Potchefstroom and location of study area and associated green areas (Source: Cilliers and Cilliers (2016))

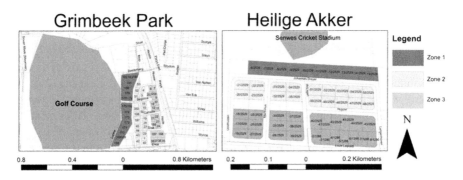

Fig. 8.4 Example of selection of zones within each area (Source: Cilliers and Cilliers (2016))

The proximity principle was rejected in this case study, as it found that residential properties located adjacent to green spaces had a lower price per square meter than properties located further away. Possible reasons for such contradictions to international findings might be related to ecosystem disservices, crime rates and noise (Konijnendijk et al. 2013:22; Perry et al. 2010). Given the unique challenges and characteristics of South African neighbourhoods, this researched called for social issues (especially *safety* considerations) and ecosystem disservices to be addressed in integrative spatial planning approaches (Cilliers 2009; Cilliers and Cilliers 2016).

8.3.2.3 Contradicting Results of Compensation Hypothesis in Rural Areas

Research conducted by Lategan and Cilliers (2017) reflected on the planning of green space and the related impact of informal backyard rental densification in South Africa, based on the compensation hypothesis. Data were retrieved from a quantitative research survey based on convenience sampling, distributed in Bridgton and Bongolethu townships, located on the outskirts of Oudtshoorn in South Africa in 2013, where 101 questionnaires were completed by residents of properties included in the demarcated area. Surveys were conducted with the assistance of chaperones, supporting the researchers in terms of points of entry to the community. Survey questions focused on respondents' access to and use of public green space, domestic green space (gardening) trends and informal backyard rental particulars, where applicable. As a convenience and not a random sample was used, p-values were reported for the sake of completeness. Findings in the Bridgton/Bongolethu case study disproved the compensation hypothesis as an assumed increase in the use of public green space in compensation for private green space lost (Lategan 2016). Over 80% of the respondents claimed to make use of proximate public green spaces, but the majority did so infrequently, not as part of their daily or even weekly routines. Statistical analysis revealed that only an insubstantial number of respondents regarded public green spaces as their children's primary play locales, with the majority still playing in domestic green spaces (private gardens), in both front and backyard spaces. The research illustrated the dissimilar functions of public and private green spaces and that such could not be provided as a substitute for the other (Haaland and Van den Bosch 2015).

8.3.2.4 Social Issues Imparting on Sustainability

As part of a qualitative inquiry into the planning of child-friendly spaces, *structured interviews were conducted in 2014* with five purposefully selected experts actively working in the Vaalharts area, a typical rural area with high vulnerability, inadequate infrastructure and basic services, located across the Northern Cape and North West Provinces of South Africa. Data of the interviews were coded and thematically analysed to provide qualitative insight on specific issues identified as problematical within this rural area. Safety issues were revealed as most crucial consideration in this context. However, in terms of the urban context, broad reference to safety issues usually implies access to and from a space, referring to restriction and division of vehicles and pedestrian spaces, and in some cases, also crime considerations addressed in terms of lighting provision or visibility measures. In rural context, it became evident that safety issues imply perceptions of crime, but more importantly it refers to restraint access to natural elements such as open water cannels where, in the Vaalharts case, it was linked to numerous drownings of children per year in the said cannels (Pienaar 2014; Kriel 2014; De Jong 2013). The research illustrated that social considerations were core in guiding spatial patterns and future development proposals.

Research of Cilliers and Rohr (forthcoming) stated how water provision and water management is considered more of a social concern than an environmental or economic concern in South Africa. This is due to the government's promise to provide free basic services for the poor (Fisher-Jeffes et al. 2012), conflating water provision with broader social equity challenges, entailing being fair and specific, providing people with what is contextually needed in order to grant equal opportunities. In this sense the community needs are related to status and access to water services, thus not planned from a sustainability perspective, but solely from a social perspective. Again, social considerations are the main driver of spatial planning in this context.

8.4 Conclusion: The Undervaluation, but Extreme Importance, of Social Sustainability

Sustainability thinking requires transdisciplinary research approaches. This is a daunting objective for cities, but even more so for cities in developing countries, still struggling with issues of illiteracy, basic service provision and increasing poverty, exacerbated by increasing urbanisation pressures and budget constraints (Cilliers and Cilliers 2015; Kuruneri-Chitepo and Shackleton 2011). This research argues that the core constraint of realising sustainability in developing countries often relates to social issues, which is ironically the least understood of the three dimensions of sustainability.

However, from a spatial planning perspective, it is evident that social issues will from now on play the leading role when considering sustainability. The balance between the natural and built environments will be negotiated in terms of liveability considerations, enforced by solid context-based planning approaches. The research emphasised the importance of context-based planning within broader social sustainability thinking, providing evidence of the unique social context in South Africa and the impact that such have on planning studies.

It was concluded that the literature-base supporting sustainable development objectives and practices in South Africa is limited, often relying on international accepted theories. Some research has recently illustrated the disparities between the international context and South African context, questioning the validity of translating international theories to local context, with specific contradictions to the proximity principle in one case study and disproval of the compensation hypothesis in another.

Adequate knowledge and contextualisation of concepts should be emphasised within a social sustainability approach. This has nothing to do with education levels or scope of training provided in developing countries but concerns the interpretation of concepts. Misunderstanding and misinterpretation of concepts used in different context, or interchangeably in various disciplines, (Escobedo et al. 2011) lead to a value gap (Cilliers 2009), where different stakeholders value a concept different, as a result of the interpretation thereof. The concept of 'green spaces' illustrated such challenge where the international accepted definition could not be applied in local context. There is no typology for open spaces that is suitable for the South African context, and as such, if communities are probed to value these open spaces, it is often perceived as having no (social, economic or environmental) value, being an abandoned open area, a crime hotspot or area demarcated for future development. The lack of definition and contextualisation is visible in practice where qualitative green spaces are perceived as a scarce commodity. The social constraint (knowledge and interpretation of the concept) are reflected in the spatial reality (lack of physical spaces).

The lack of adequate context-based research further constraints this problem. The actual value and benefits that green spaces might provide to South African communities, both urban and rural, should be explored and translated to a monetary value, to substantiate the motivation thereof and build a case in favour of sustainability planning. Methods, theories and equations of urban economics and green economics should be translated into urban planning approaches to inform decision-making (Bertaud 2010:1; Luttik 2000:161, 162). It is within this structure that location-specific issues should be included, such as cultural preferences related to status (enclosed versus open spaces; access to services), and safety issues (actual safety versus perceived safety, for different stakeholders and communities).

Finally the issue of scale cannot be ignored when considering context-based planning. The majority of sustainable thinking and related theories refer to broader environmental processes at a regional scale that have not been translated more practically to a local government level tasked with implementation (Cilliers and Cilliers 2016). This holds a great challenge for South African cities which are often defined by smaller administrative boundaries. Sustainability thinking should thus go beyond discussions on intergovernmental cooperation to enforce ground level implementation, engraved in social considerations.

Acknowledgements The financial assistance of the National Research Foundation (NRF) towards this research is hereby acknowledged. Opinions expressed and conclusions arrived at are those of the authors and are not necessarily to be attributed to the NRF.

Bibliography

Basiago, A. D. (1999). Economic, social, and environmental sustainability in development theory and urban planning practice. *The Environmentalist, 19*, 145–161.

Beck, H. (2009). Linking the quality of public spaces to quality of life. *Journal of Place Management and Development, 2*(3), 240–248.

Bedimo-Rung, A. L., Mowen, A. J., & Cohen, D. A. (2005). The significance of parks to physical activity and public health: A conceptual model. *American Journal of Preventive Medicine, 28*(2S2), 159–168.

Bertaud, A. (2010). The study of urban spatial structures. http://alain-bertaud.com. Accessed 4 July 2010.

Bolitzer, B., & Netusil, N. R. (2000). The impact of open spaces on property values in Portland, Oregon. *Journal of Environmental Management, 59*(3), 185–193.

Bolund, P., & Hunhammar, S. (1999). Ecosystem services in urban areas. *Ecological Economics, 29*(2), 293–301.

Brander, L. M., & Koetse, M. J. (2011). The value of urban open space: Meta-analyses of contingent valuation and hedonic pricing results. *Journal of Environmental Management, 92*(10), 2763–2773.

CABE Space. (2005). *Paying for parks: Eight models for funding urban green space.* London: Commission for Architecture and the Build Environment.

Carrasco, M. A., & Bilal, U. (2016). A sign of the times: To have or to be? Social capital or social cohesion? *Social Science & Medicine, 159*, 127–131.

Caspersen, O. H., Konijnendijk, C. C., & Olafsson, A. S. (2006). Green space planning and land use: An assessment of urban regional and green structure planning in greater Copenhagen. *Geografisk Tidsskrift, Danish Journal of Geography, 106*(2), 7–20.

Chen, W. Y., & Jim, C. Y. (2010). Amenities and disamenities: A hedonic analysis of the heterogeneous urban landscape in Shenzhen (China). *Geographical Journal, 176*(3), 227–240.

Chiesura, A. (2004). The role of urban parks for the sustainable city. *Landscape and Urban Planning, 68*(1), 129–138.

Cilliers, S. S., Siebert, S. J., Davoren, E., & Lubbe, C. S. (2012). Social aspects of urban ecology in developing countries, with an emphasis on urban domestic gardens. In M. Richter & U. Weiland (Eds.), *Applied urban ecology: A global framework* (pp. 123–138). Chichester: Wiley-Blackwell.

Cilliers, E. J., & Cilliers, S. S. (2015). From green to gold: A South African example of valuing urban green spaces in some residential areas in Potchefstroom. *Town Planning Review, 67*, 1–12.

Cilliers, E. J., & Cilliers, S. S. (2016). *Planning for green infrastructure: Options for South African cities.* Johannesburg: South African Cities Network.

Cilliers, E. J., & Rohr, H. E. (Forthcoming). Integrating WSUD and mainstream spatial planning approaches: Lessons from South Africa. Chapter 23.

Cilliers, E. J. (2009). Bridging the green-value-gap: A South African approach. *International Journal of Environmental, Chemical, Ecological, Geological and Geophysical Engineering, 3*(6), 182–187.

Cilliers, E. J. (2010). *Rethinking sustainable development: The economic value of green spaces.* Dissertation for completion of M.Com Economics, Potchefstroom: North West University.

Cilliers, E. J., Diemont, E., Stobbelaar, D. J., & Timmermans, W. (2010). Sustainable green urban planning: The green credit tool. *Journal of Place Management and Development, 3*(1), 57–66.

Cilliers, S. S., Cilliers, E. J., LUBBE, R., & SIEBERT, S. (2013). Ecosystem services of urban green spaces in African countries — Perspectives and challenges. *Urban Ecosystems, 16*(4), 681–702.

Cities Alliance. (2007). *Liveable cities: The benefits of urban environmental planning. A cities alliance study on good practices and useful tools* (p. 162). Washington: York Graphic Services.

Crompton, J. L. (2001). The impact of parks on property values: A review of the empirical evidence. *Journal of Leisure Research, 33*(1), 1–31.

De Jong, N. (2013). *Addressing social issues in rural communities by planning for lively places and green spaces.* Dissertation submitted to the North-West University, Potchefstroom, 2013.

De Wit, M. P., & Blignaut, J. N. (2006). *Monetary valuation of the grasslands in South Africa making the case for the value of ecosystem goods and services in the grassland biome.* Report prepared for Lala Steyn at South African National Biodiversity Institute.

Dixon, T., & Woodcraft, S. (2013). Creating strong communities- measuring social sustainability in new housing development. *Town and Country Planning*, 473–480.

Escobedo, F. J., Kroeger, T., & Wagner, J. E. (2011). Urban forests and pollution mitigation: Analyzing ecosystem services and disservices. *Environmental Pollution, 159*(8–9), 2078–2087.

EU European Union. (2013). *Building a green infrastructure for Europe*. Luxembourg: Publications Office of the European Union.

Fausold, C. J., & Lilieholm, R. (1999). The economic value of open space: A review and synthesis. *Environmental Management, 23*(3), 307–320.

Fisher-Jeffes, L. N., Carden, K., Armitage, N. P., Spiegel, A., Winter, K., & Ashley, R. (2012). *Challenges facing implementation of water sensitive urban design in South Africa*. Proceedings of the 7th International Conference on Water Sensitive Urban Design, Melbourne, Australia.

Forman, T. T. (2013) Ecological resilience as a foundation for urban design and sustainability. In S. T. A. Pickett, M. L. Cadenasso, & B. McGrath (Eds.), *Resilience in ecology and urban design*. Dordrecht Heidelberg New York London: Springer, New York.

Godschalk, D. R. (2004). Land use planning challenges: Coping with conflicts in visions of sustainable development and livable communities. *Journal of the American Planning Association, 70*(1), 5–13.

Goel, S., & Sivam, A. (2014). Social dimensions in the sustainability debate: The impact of social behaviour in choosing sustainable practices in daily life. *International Journal of Urban Sustainable Development, 7*(1), 61–71.

Haaland, C., & Van den Bosch, C. K. (2015). Review: Challenges and strategies for urban green-space planning in cities undergoing densification: A review. *Urban Forestry & Urban Greening, 14*(4), 760–771.

Hansmann, R., Hug, S. M., & Seeland, K. (2007). Restoration and stress relief through physical activities in forests and parks. *Urban Forestry & Urban Greening, 6*(2007), 213–225.

Hardin, B. (2001). Case study using market price methods: Estimating the value of ecosystem functions using the replacement cost method. In Turpieet et al. (Eds), *Valuation of open space in the cape metropolitan area. A valuation of open space in the cape metropolitan area*. Report to the City of Cape Town.

Herzele, A., & Wiedemann, T. (2002). A monitoring tool for the provision of accessible and attractive urban green spaces. *Landscape and Urban Planning, 63*(2), 109–126.

Hodgkison, S., & Hero, J. M. (2002). The efficacy of small-scale conservation efforts, as assessed on Australian golf courses. *Biological Conservation, 135*(4), 576–586.

Huston, G. D. (2016). *Evaluating local green infrastructure training and education approaches within urban planning curricula*. Mini-dissertation submitted in partial fulfilment of the requirements for the degree Baccalareus Artium et Scientiae in Urban and Regional Planning at the Potchefstroom Campus of the North- West University. Potchefstroom.

Irwin, E. G. (2002). The effects of open space on residential property values. *Land Economics, 78*, 465–480.

Kazmierczak, A. E., & James, P. (2008). *The role of urban green spaces in improving social inclusion*. Salford: University of Salford, School of Environment and Life Sciences.

Kong, F., Yin, H., & Nakagoshi, N. (2007). Using GIS and landscape metrics in the hedonic price modeling of the amenity value of urban green space: A case study in Jinan City, China. *Landscape and Urban Planning, 79*(3–4), 240–252.

Konijnendijk, C. C., Annerstedt, M., Nielsen, A. B., & Maruthaveeran, S. (2013). *Benefits of urban parks: A systematic review*. A Report for IFPRA, Copenhagen & Alnarp, January 2013.

Kriel, M. (2014). *Planning child-friendly spaces for rural areas in South Africa: The Vaalharts case study*. Dissertation submitted to the North-West University, Potchefstroom. 2014.

Kuo, F. E. (2003). The role of arboriculture in a healthy social ecology. *Journal of Arboriculture, 29*(3), 148–155.

Kuruneri-Chitepo, C., & Shackleton, C. M. (2011). The distribution, abundance and composition of street trees in selected towns of the eastern cape, South Africa. *Urban Forestry & Urban Greening, 10*(3), 247–254.

Landscape Institute. (2013). Green infrastructure: An integrated approach to land use. London. Available at http://www.landscapeinstitute.org/PDF/Contribute/2013GreenInfrastructureLIPositionStatement.pdf. Date of access: 22 Mar 2016.

Lategan, L. G. (2016). *Reflecting on South Africa's informal backyard rental sector from a planning perspective*. Ph.D thesis at the North-West University, South Africa.

Lategan, L. G., & Cilliers, E. J. (2017). Considering urban green space and informal backyard rentals in South Africa: Disproving the compensation hypothesis. *Town and Regional Planning, 69*, 1–16.

Luttik, J. (2000). The value of trees, water and open space as reflected by house prices in the Netherlands. *Landscape Urban Planning, 48*(3), 161–167.

McConnachie, M., & Shackleton, C. M. (2010). Public green space inequality in small towns in South Africa. *Habitat International, 34*, 244–248.

McPherson, E. G., Maco, S. E., Simpson, J. R., Peper, P. J., Xiao, Q., Van Der Zanden, A. M., & Bell, N. (2002). *Western Washington and Oregon community tree guide: Benefits, costs, and strategic planning*. Silverton: International Society of Arboriculture.

Natural Economy North West. (2007). *The economic value of green infrastructure*. North West England. 20p.

Perman, R., Ma, Y., McGilvray, J., & Common, M. (2003). *Natural resource and environmental economics*. Harlow: Pearson Education.

Perry, E. D., Moodley, V. & Bob, U. (2010). *Open spaces, nature and perceptions of safety in South Africa: A case study of Reservoir Hills, Durban*. School of Environmental Science, University of KwaZulu-Natal. 17p.

Pienaar, A. (2014). *Structured interview*. Potchefstroom.

Richmond, B. (1993). Systems thinking: critical thinking skills for the 1990s and beyond. *System dynamics review, 9*(2), 113–133.

Riddel, R. (2004). *Sustainable urban planning: Tipping the balance*. Blackwell Publishing Ltd..

Roberts, D. C., Boon, R., Croucamp, P., & Mander, M. (2005). Resource economics as a tool for open space planning Durban, South Africa. In: T. Trzyna (Ed.), *The Urban Imperative, urban outreach strategies for protected area agencies*. Published for IUCN-California Institute of Public Affairs (pp. 44–48). IUCN, Sacramento: California Institute of Public Affairs.

Roger, S.U. (2003). *Health benefits of gardens in hospitals: Plants for People*. Texas: Centre for health systems and design.

Schäffler, A., Christopher, N., Bobbins, K., Otto, E., Nhlozi, M. W., De Wit, M., Van Zyl, H., Crookes, D., Gotz, G., Trangoš, G., Wray, C., & Phasha P. (2013). State of Green Infrastructure in the Gauteng City-Region. Gauteng City-Region Observatory (GCRO), a partnership of the University of Johannesburg, the University of the Witwatersrand, Johannesburg, and the Gauteng Provincial Government.

Schäffler, A., & Swilling, M. (2013). Valuing green infrastructure in an urban environment under pressure — The Johannesburg case. *Ecological Economics, 86*(2013), 246–257.

Stigsdotter, U. A. (2008). *Urban green spaces: Promoting health through city planning*. Sweden: Swedish university of agricultural sciences. 17p.

Stiles, R. (2006, December). *Urban spaces – enhancing the attractiveness and quality of the urban environment*. WP3 Joint Strategy. University of Technology, Vienna.

Sutton, C. M. (2006). *On urban open space: A case study of Msunduzi Municipality, South Africa*. Canada: Queens University. (Thesis – B.Sc). School of Environmental Studies. 139 p.

Swanwick, C., Dunnett, N., & Woolley, H. (2003). Nature, role and value of green space in towns and cities: An overview. *Built Environment, 29*(2), 94–106.

Taylor, A. F., Kuo, F. E., & Sullivan, W. C. (2002). Views of nature and self-discipline: Evidence from inner city children. *Journal of Environmental Psychology, 22*, 49–63.

Thomas, K., & Littlewood, S. (2010). From green belts to green infrastructure? The evolution of a new concept in the emerging soft governance of spatial strategies. *Planning, Practice & Research, 25*(2), 203–222.

Tlokwe City Council. (2010). Tlokwe City Council Valuation Roll for the period 2009/2013. Potchefstroom.

Turpie, J., & Joubert, A. (2001). Case studies using revealed preference methods I: Estimating the recreational use value of Zandvlei using the travel cost method. In Turpie et al. (Eds.), *Valuation of open space in the cape metropolitan area*. A valuation of open space in the Cape Metropolitan Area, Report to the City of Cape Town.

Tyrvainen, L. (1997). The amenity value of the urban forest: An application of the hedonic pricing method. *Landscape and Urban Planning, 37*(3–4), 211–222.

Ulrich, R. S., & Addoms, D. L. (1991). Psychological and recreation benefits of a Recreational Park. *Journal of Leisure Research, 13*(1), 43–65.

United Nations. (2017). Progress towards the sustainable development goals. Report to the Secretary-General. E/2017/66. Available at: http://www.un.org/ga/search/view_doc.asp?symbol=E/2017/66&Lang=E. Date of access 5 July 2017.

Van den Berg, A., Hartig, T., & Staats, H. (2007). Preference for nature in urbanized societies: Stress, restoration, and the pursuit of sustainability. *Journal of Social Issues, 63*(1), 79–96.

Van Leeuwen, E., Nijkamp, P., & de Norohna Vaz, T. (2009). *The multi-functional use of urban green space.* Amsterdam. Faculteit der Economische Wetenschappen en Bedrijfskunde Research Memorandum, (2009-51):1–13

Veiga, R. S. (2015). *A proposed green planning development framework: Integration of spatial planning and green infrastructure planning approaches.* Dissertation submitted in fulfilment of the requirements for the degree Magister Artium et Scientiae in Urban and Regional Planning at the Potchefstroom Campus of the North- West University. Potchefstroom.

WCED World Commission on Environment and Development. (1987). *Our common future.* Oxford: Oxford University Press.

Woolley, H., Swanwick, C., & Dunnet, N. (2003). Nature, role and value of green space in towns and cities: an overview. www.atypom-link.com/ALEX/doi/abs/10.2148/benv.29.2.94.54467. Accessed 18 Sept 2009.

Wright, H. (2011). Understanding green infrastructure: The development of a contested concept in England. *Local Environment, 16*(10), 1003–1109.

Chapter 9
Smart Eco-Cities Are Managing Information Flows in an Integrated Way: The Example of Water, Electricity and Solid Waste

Meine Pieter van Dijk

9.1 Introduction

Smart cities are cities which manage their flows of waste, energy, people, goods and services in an integrated way. We all want to be smart and everything has to be smart. The benefits of attending a conference on shaping the urban future in Manchester were advertised in 2016 as manifold: explore innovative ways to use technology and data to make urban life safer, smarter and more sustainable. You would hear from those initiating and driving urban change towards a smarter city how to implement smart technologies to engage with energy, transport and other data. You would learn to explore new practices, new thinking and to obtain new contacts which would support you taking the next steps towards a more efficient city. Finally, you would network with organizations from across the globe to explore more efficient ways for the smart city revolution. The expectations are high.

In this chapter, we will first discuss the urban issues at stake and what a smart city is, but also provide some critical remarks about the smart city concept and compares it to concepts such as resilient and eco-cities. The following sections look at what this data processing-based approach to urban management means for urban water management and solid waste collection. Many of the examples are chosen from our research in China and Europe.

M. P. van Dijk (✉)
Maastricht School of Management (MSM), Maastricht, Netherlands

International Institute of Social Studies (ISS) of Erasmus University Rotterdam, The Hague, Netherlands

Beijing University of Civil engineering and Architecture (BUCEA), Beijing, China
e-mail: DijkM@msm.nl

© Springer International Publishing AG, part of Springer Nature 2018   149
M. Dastbaz et al. (eds.), *Smart Futures, Challenges of Urbanisation, and Social Sustainability*, https://doi.org/10.1007/978-3-319-74549-7_9

9.2   What Are the Issues?

Urban management starts with identifying the issues, preferably in a participatory way. We can classify the issues as economic, social or environmental and will give some examples. The Dutch association of municipalities (Binnenlands Bestuur 2017a: 53) formulates as the major challenges for cities: constructing enough houses, the energy transition, investments in climate change adaptation and the digital city. At the same time, cities have to be competitive, pleasant places to live in, and they have to be resilient against all kinds of crises and disasters.

The problems of Third World cities are well known and are often larger than problems of cities in developed countries. This is not just because of the size of the population and a more rapidly growing urban population but also because of a smaller formal sector, which usually generates fewer funds for local governments while a large section of the population is living below the poverty line. Our definition in managing cities smartly is using information technology in urban management (Van Dijk 2006: 123).[1] We will give examples of both developed and developing cities showing how they deal with these issues using modern information technology.

A different approach to these problems may be possible using the now available information technology (IT) and urban information systems (UIS). IT allows us to combine different information systems. Integrated analysis of urban issues, which is then possible, can help to improve urban management and hence reduce urban problems. Utilizing available databases in a proficient way in the urban areas may be a challenge.

Economic issues concern making your city a dynamic city. It means dealing with urban poverty, unemployment and the lack of competitiveness of certain cities, which may not manage to attract investment or renovate their centre.

Economic stagnation and recession may lead to all kinds of social issues, affecting certain groups more than others. The dichotomy is usually between the poor and the rich, the migrants and the people born in the city, the people with education and the people with no or very limited education. It is the lower classes versus the newly emerging urban middle class and the government versus private sector representatives.

Finally, we want a sustainable city, meaning urban managers have to deal with environmental issues, including pollution, but also the consequences of climate change, which may imply flooding (Jakarta), drought (Northern China), or higher temperatures (all cities). There exist possible adaptation mechanisms in urban areas, but the question is to what extent do urban actors pursue these strategies. In the Netherlands, the bigger cities have plans to become more climate proof (e.g. Rotterdam 2008), but the smaller municipalities have a hard time to catch up. This

---

[1] The European Union (EU) has the Asia Urbs programme which provides digital links to best practice databases on, for example, ecological urban renewal, local European local transport information systems and other best practices. The EU Horizon 2020 programme also offers funding for smart cities' lighthouse projects (editor@smartcitiesconnect.org).

is one of the reasons to also develop at the national level a Delta plan for spatial adaptation (Binnenlands Bestuur 2017b).

In this chapter, we will deal in particular with the economic and environmental issues. The social issues not dealt with concern inequality, segregation, the shrinking of the middle class or consequences of globalization, etc.

9.3   What Are Smart Cities?

You may be wondering what smart cities are? Townsend's (2013) book on smart cities provides 'An unflinching look at the aspiring city-builders of our smart, mobile, connected future'. In his words: 'We live in a world defined by urbanization and digital ubiquity, where mobile broadband connections outnumber fixed ones, machines dominate a new "internet of things," and more people live in cities than in the countryside'.

What is specific about smart cities? There are at least three types of definitions of smart cities, definitions which may help to understand the urban metabolism, but also show that people do not always have the same thing in mind when they talk about smart cities:

A. Emphasizing the use of modern technology, for example, the use of Internet of things for all kinds of problems (applied in a neighbourhood of Seoul, the capital of South Korea) and the importance of innovation for the city of the future (Box 9.1).
B. Stressing transition and adaptation to rapid changes, by pooling knowledge, sharing best practices and considering different initiatives to tackle challenges faced. This is the innovative planning approach used to create the cities of the future introduced by transition managers who want to increase our capability to adapt to rapid changes (Van der Steen et al. 2010).
C. Pointing to different ways of managing cities, focusing on managing flows in cities and using information and communication technologies (ICT) and geographic information systems (GIS) to do that in an integrated way; this is the definition which I will use in the rest of this chapter.

Box 9.1 The Internet of Things (IoT): Some Examples
- Receiving warnings on your phone or wearable device when IoT networks detect some physical danger is detected nearby
- Self-parking automobiles
- Automatic ordering of groceries and other home supplies
- Automatic tracking of exercise habits and other day-to-day personal activity including goal tracking and regular progress reports.
- Location tracking for individual pieces of manufacturing inventory

Source: Shrouf and Miragliotta (2015).

The IoT can be defined as the interconnection of physical objects and other physical objects to serve the interests of human beings, animals or the earth in general.[2] The IoT is about physical objects with an intelligent interface to communicate with other objects, users and environments. It will transform our environment in many ways, including the ways we use energy, water and other resources within the 'smart cities' of the future.

I opt for the third definition. It means cities are no longer considered a collection of houses and roads, but rather the sum of a number of flows that need to be managed. Important flows which will be discussed briefly are the water cycle (including drinking and waste water), the waste and the energy cycle (including the reduction of the reduction of CO_2 and other greenhouse gas emissions). Flow of goods and the resulting waste flows require waste minimization and integrated waste management. Mobility is a flow of people and goods and requires the development of integrated infrastructure and transport policies, etc. Sensors can be put at critical points (in bridges, traffic lights and parks) and digitally inform the authorities about air pollution (Qui and Van Dijk 2015), noise pollution or concentrations of toxic substances in surface water.

9.4   Managing Urban Information Flows

In Van Dijk (2006) a whole chapter is dedicated to the use of information technology in urban management. A decade ago, the challenge was formulated as:

> To design an integrated urban information system (UIS), which allows urban decision makers to preview the consequences of their decisions and to monitor the results of implemented policies and investments. Such an integrated UIS will not only represent a powerful educational tool, but will also be of great practical value for urban managers.

It is emphasized that: 'access to digital data allows more transparency and more participatory decision making in urban affairs by inhabitants and other urban actors. Computers also facilitate the provision of training and information to urban managers and the population at large'.

Examples of given applications are: 'a (digital) land registration system as a basis for urban planning, property tax and housing policies. Another application could be related to taxing in general. Computers may be used to identify the assets of an inhabitant and tax him or her accordingly. Applications could also concern the monitoring of the spread of certain diseases, the effects of certain policies on target groups or the impact of credit programs on the development of economic activities by poor people'. See also Box 9.2.

---

[2]This is my definition; putting it in general terms helps to emphasize the potential: warning and action on pollution, filling your fridge and milking your cows!

Box 9.2 Applications of Information Technology in Urban Management

1. Urban planning: planning becomes easier by using computers, but computers also allow the planners to use geographical and socioeconomic or demographic data and combine these data through GIS.

2. Social development policies: the use of IT and IS allows cities to define and implement social development policies tailored to the needs of the population and to monitor their effects.

3. Land information systems can become tools of urban management if also used for planning and social policy purposes. Computers allow an objective and transparent registration of land ownership and changes in it.

4. Participation can be promoted through IT: increased participation is possible through new technological opportunities. People can use the Internet to collect information and provide feedback or household data to local authorities.

5. Monitoring locational changes: combining GIS software with data on the movement of people/businesses allows municipal authorities to monitor locational changes and to assess their urban management consequences.

6. Dealing with urban congestion becomes easier if the transport department has good IS about the traffic movements and instruments to influence them.

7. Monitoring environmental developments: data on air and water pollution can be transferred easily to a central computer, which can warn if certain thresholds are passed.

8. Other urban management applications: a number of applications, which can only be mentioned. For example, using the Internet to inform small entrepreneurs about new technologies/products.

9. Single sector-specific applications: in practice computers and data sets are often used for simple management purposes, such as registering the taxes or keeping track of the municipal finances.

10. Capacity building: computers provide great opportunities to train people and using computers effectively requires a lot of training.
    Source: Van Dijk (2006).

Paulsson (1992) studied already the use of satellites, remote sensing and GIS analysis for urban planning in the Third World.[3] The use of IT and UIS in, for example, Cebu and Dakar, showed different but very promising applications. It can be concluded that IT and UIS can be used for dealing with a range of urban issues, such as unemployment, poverty, participation, supplying urban services, etc. The challenge is to assess how existing data from larger IS can be combined for this purpose. A large-scale introduction of IT in the administrative, educational and

---

[3] Other examples of the use of GIS in urban planning have been studied by de Bruijn (1987) and Turkstra (1998).

training systems of developing countries will most likely translate into a boost to their economies. It will create jobs, may open new markets and can lead to marked cost reductions in fields such as education and training. The necessary investments may be substantial, but if properly used, the gains would also be important. Sometimes it is just a question of using existing data differently, which may now be possible because of the availability of more powerful computers.

This question can be raised: How much experience do we have with managing urban information flows now? Two important flows are the water cycle and the energy cycle. In this definition, smart cities are focusing on managing these flows. The challenge is to achieve integration in the framework of urban management, using modern information technology.

The easy examples concern utilities providing water or electricity. These examples immediately present some problems: how do we deal with mixing the different sources of energy (wind, solar, conventional) in a rational way, trying at the same time to reduce CO_2 and other greenhouse gas emissions? In the water sector, we would also like a smart and a more ecological approach. For many cities facing climate change and water shortages, it means they should implement eco adaptive or integrated water resources management, which may mean trying to close the water cycle and focus on the most important issues together with the relevant stakeholders. However, like in the case of energy, the challenge is also to use alternative technologies such as rainwater harvesting (Liang and van Dijk 2018), aquifer refill and water demand management, to increase the supply of water.

What can we expect in the cities due to climate change, knowing the sectors affected, in particular water and energy:

* High temperatures, more rain, less rain more volatility
* Necessary: to store water
* No to provide building permits for lower-lying areas
* Rainwater harvesting and separating grey and brown water
* Importance of the different levels of government, in particular, the household level

To make it more complicated, the different flows may interact. Treating your waste water in a closed water cycle may require energy, but also generates energy that you need to consider in addition to your existing energy flows. Plus, I have added the criterion sustainability. My definition of a sustainable or an eco-city is one that emphasizes that it is accessible to everyone. It is a city in balance with nature; a city that reduces, recycles and reuses waste; and a city that has closed its water cycle and is integrated into the surrounding region. The combination, smart green or smart eco-cities, requires different, more integrated or collaborating urban governance structures dealing with different sectors or issues, which are participatory and self-learning. There should be a division between different modes of governance: hierarchy, the market and collaboration. This requires leadership, an urban manager may play this role in my perspective: the mayor in many cities or the municipal commissioner in most Indian cities.

These ideas had serious consequences for the definition of urban management which I use in my teaching and consultancies. In my book *Managing Cities in Developing Countries* in 2006, it was still simple. Urban management is 'the effort to coordinate and integrate public as well as private actions to tackle the major problems inhabitants of cities are facing in an integrated way, to make a more competitive, equitable and sustainable city'. If this is too long for you, the short definition is: urban management is putting a plan into practice! When you go for managing flows, a new definition is required. If urban management is based on managing flows it means 'getting the food into the city and the shit out of the city, and preferably in a sustainable way!'

9.5 Critique of the Smart City Approach

Kitchin (2015: 1) formulates four points of critique on the rhetoric of smart cities, which according to him appear as nonideological, commonsensical and pragmatic. The four shortcomings are as follows:

1. The lack of detailed genealogies of the concept and initiatives
2. The use of canonical examples and one-size-fits-all narratives
3. An absence of in-depth empirical case studies of specific smart cities initiatives and of comparative research that contrasts smart city developments in different locales
4. Weak engagement with various stakeholders

We can add that using big data analytics and cloud-based services in managing smart city initiatives is also expanding market for control and will rise all kinds of ethical issues, like do we want the government to know all the time where we are and what we are doing? The *Financial Times* (13-8-2017) dedicated its House & Home section to the issue of 'giving up our data and domestic lives to machines'. Everything becomes easier, yet: 'the other side of delegation is loss of control'.

9.6 Other Typologies of Cities

Urban governance structures are challenged because of local environmental issues and urgent events such as urban floods and drinking water shortages (Van Dijk et al. 2017). Many different ways of classifying cities can be found in the literature.

Table 9.1 summarizes some of the best known ones. Given we discussed already the smart city concept, we will now go into more detail for the eco-city concept, the resilient city concept, the sponge city concept and the smart eco-city idea.

Table 9.1 Classifying cities: some relevant concepts.

Type of city/reference	Major characteristic
Climate-proof cities (Van Dijk 2014a, b; Rotterdam 2008)	Focus is often on reduction of GHG emissions and introducing resilience
Cultural city (Florida 2004)	Cities no longer places of manufacturing but of 'cultural' activities, which are important in the future
Eco-city (Van Dijk 2015)	Different dimensions are suggested for defining ecological cities, such as their concern about the quality of urban life
Resilient cities? (Rockefeller foundation 2014)	Assessing the ability to learn, plan and recover from the hazards to which these cities are exposed
Smart city (Hajer and Dassen 2014)	Smart eco-cities are managing flows of information in an integrated way
Sponge city concept (source water magazine 2017)	Efforts to slow, spread, sink and store runoff (mainly deals with flooding)
Sustainable or green cities (UN Habitat 2014)	A city that reduces, recycles and reuses waste and a city that has closed its water cycle and deals with pollution

9.7 The Ecological City of the Future

In Beijing, there are about 30,000 ecological initiatives under way to make the city greener and more sustainable. To assess their governance structure and ecological character, Liang and Van Dijk (2012) reviewed how Beijing and its inhabitants deal with water and sanitation issues and climate change, as well as looking at pollution, energy consumption, solid waste, transport and housing.

What would the ecological city of the future look like? The eco-city of the future is not just about dealing with environmental issues. Such a city will also need a sound economic basis; appropriate solutions for its transport systems, for housing and for urban service delivery; and the provision of urban amenities. There is more and more attention for the quality of life in cities, and the presence of sufficient urban services and amenities is an important factor to make a city attractive.

The emphasis is on the experiences with adaptation to climate change in urban China. In the discussion about good governance, which was introduced by donors and donor organizations, there is often a cultural undertone. What is considered poor governance or corruption in one country according to certain norms and values may be considered as providing incentives in another country. However, the need for good governance and more transparency is often stressed to increase the chances of success of urban projects.

What are the eco-city initiatives initiated by the government, and are they actually working through at the local level? Li et al. (2013) studied how Chinese institutions are developing policies to help local communities in Lanchang river basin to cope with problems of climate variability and change; they are mapping the relevant policies and interviewing stakeholders to assess how to improve drought preparedness. According to the constitution, the national laws made by the National People's Congress (or its standing committee) set the principles, for example, for dealing with climate change. The national and local ordinances, which are passed by the state council or local councils, are used to work out the details. Subsequently, the ministry or local government concerned designs its own rules (according to its responsibilities) to implement law and ordinance. Governance is embedded in a larger system of regional, provincial, municipal and district level government structures (Hou 2000).

Chinese cities have introduced all kinds of policies and programmes to deal with issues like climate change and pollution. Many cities want to become more ecologically friendly, or eco-cities. Initiatives taken to become a more sustainable city or to create ecological neighbourhoods or eco-cities have been documented. Elements that recur are closing the water cycle, reducing greenhouse gas emissions, opting for waste minimization and integrated waste management and developing integrated infrastructure and transport policies. How are these initiatives governed? Several examples will be given. The challenge is to achieve integration in the framework of smart urban management, coordinating the efforts made by different stakeholders to solve the issues in an integrated way.

9.8 The Sponge City Concept

The Chinese have developed the sponge city concept, to slow, spread, sink and store runoff.[4] Many cities are improving the runoff of water and the capacity to store the rainwater temporarily. They created water absorption beds, for example, in Tianjin city. Infiltration by design helps recharge urban aquifers, mitigate floods and let city surfaces breathe. Cities choose between cisterns, rooftop gardens, retention ponds and permeable pavements to reduce half to nearly all runoff. The efforts are known in the Switch project (Van der Steen et al. 2010) and can be called water-sensitive urban design (WSUD in Australia), low impact development (LID in North America) or sustainable urban drainage systems (SUDS in Europe).

9.9 Resilient Cities

De Jong (2016) defines urban resilience as 'The ability to learn, plan, and recover from the hazards to which they are exposed'. We can make the definition of resilience more complex by adding that such a city should also be able to deal with these issues in a sustainable way.[5] We need to develop an analytical framework allowing us to actually quantify what the major actors are doing. The major actors are at dif-

[4] www.thesourcemagazine.org/sponge-cities-can-chinas-model-go-global/?utm_source=IWA-NETWORK&utm_campaign=16a2d396c8-The_Source_newsletter_28_Source_list_30_08_2017&utm_medium=email&utm_term=0_c457ab9803-16a2d396c8-158952849

[5] BUCEA has a programme to study what 30 Chinese cities are doing in the field of climate change.

ferent geographical levels: the national, provincial, municipal and district level, the neighbourhood initiatives, the project developers and the individual households who may take measures to prepare themselves. Furthermore, it is important to distinguish different types of policies to achieve resilience: general policies, subsidies, planning requirements, demand management policies and other activities.

Let me give some examples of each category:

1. General policies, at the national level create a market for CO_2 emission trading rights and at the municipal level improve urban drainage
2. Subsidies, at the national level to raise awareness for the issues; subsidies for adapting buildings (e.g. isolation) and for separating grey and brown water. At the municipal level for separating grey and brown water
3. Planning requirements, at the national level low carbon electricity generation and at the local level requiring double windows or installing demo plants
4. Demand management policies, at the national level for water and energy and at the local level by making available the necessary devices

Some initiatives are taken even at lower levels, for example, at the neighbourhood level: information dissemination activities to enable these initiatives. It helps if the governance structure is decentralized. Examples are energy-related activities such as installing isolation, providing joint energy projects, using the temperature of groundwater (groundwater temperature exchange systems), etc. Different actors can play a role; for example, project developers can provide or receive subsidies. Citizens can be promoted to engage in rainwater harvesting projects or build facilities for the separation of grey and brown water. Finally, it is possible to introduce demand management to limit the consumption of water and electricity. We will now deal with managing important flows, the flow of water and the flow of energy.

9.10   Smart Eco-Cities

Van Dijk (2017) argues in favour of smart eco-cities. Cities are considered no longer a collection of houses and roads, but rather the sum of a number of flows that need to be managed. Important flows are the water cycle, the waste and the energy cycle (including the reduction of CO_2 and other greenhouse gas emissions). In this definition, smart cities are focusing on managing these flows. The challenge is to achieve integration in the framework of urban management, using modern information technology.

A lot more is possible within existing urban systems if one is ready to think in terms of creative solutions, using experiences gained elsewhere. The challenge is to introduce environmentally and financially sustainable solutions. Beijing is unique in the world because it has a number of green or eco initiatives, for example, legislation forcing all major new buildings to separate brown and grey water and to treat their grey water on the spot. The success of this policy is limited, however, since it is cheaper to buy clean municipal water than to make the effort of cleaning grey water and then using it for flushing the toilets and irrigating the garden (Liang and Van Dijk 2009).

Drivers such as rapid economic growth, urbanization, decentralization, climate change and a growing awareness of environmental degradation have created an enabling environment for citizens and entrepreneurs to take more initiative. An analysis of the multilevel governance structures in place has been carried out to determine the role of local governance structures and initiatives taken at the neighbourhood, building and household level to achieve more sustainable urban development. Citizens are taking the initiative to deal with the many challenges they are facing, which are partially climate change-related, in particular, the increasing incidence of floods in cities like Beijing and Wuhan. The research reviews the various reactions to climate change through a survey, followed by case studies in one or two of China's most affected provinces (Li et al. 2013).

In Chinese cities, more initiatives are developing at the household and enterprise level to deal with the consequences of climate change. Making Beijing ready for the consequences of climate change is an issue of governance. Climate change and a growing awareness of environmental degradation have helped to create a situation in which people have become more aware of their physical environment. Owing to climate change, we note a growing awareness that a different approach is needed from the dominant top-down and command-and-control approach to environmental issues. More specifically, we need to know:

1. How are climate change policies and programmes implemented at the city level?
2. Which governance structures have been adopted by China to implement these policies and programmes?[6]
3. What are the effects on citizens and at the enterprise level?
4. Which initiatives do they take themselves to deal with the consequences of climate change, in particular, with floods?

A large number of initiatives to make Beijing a more ecological city have been launched, ranging from separating grey and brown water to financing sophisticated ecological projects in the framework of the Olympics in 2008 (Van Dijk 2009).[7]

---

[6]Water governance has been reformed to achieve more integration and promote experiments to deal with water scarcity.

[7]The paper gives some examples of problems in the urban sector in China. Subsequently the hierarchy of Chinese government is explained, before describing the stakeholders in urban governance. A separate section is devoted to changes in the urban water governance structure that have already been implemented in Beijing.

Table 9.2 Smart solutions for urban water issues in China

Issue	Smart solution	Reference
Dealing with waste water	Separating grey and brown water and treating them differently	Liang and van Dijk (2009)
Water shortages	Rainwater harvesting and aquifer infiltration techniques	Liang and van Dijk (2011)
Too much water	Improved water management with sensors	Switch project (Van der Steen et al. 2010)
The issue of rainfall runoffs	Sustainable urban drainage systems (SUDS)	Switch project (Van der Steen et al. 2010)
Governance of urban water issues	Urban management using information technology	Liang and Van Dijk (2012) Qiu and van Dijk (2014)

9.11   Experiences with Managing Specific Flows: Water

Cities are threatened by pollution, climate change and rapid urbanization. For the water sector, this means getting enough clean water into the city and to do something with the waste water which flows out of the city. If that is not enough, the atmosphere may bring too much or not enough water. In the case of climate change, it is also predicted that rain will be more volatile. How do we deal with these issues? I will give some examples from the Switch project, which received EU funding to worry about the city of the future (Van der Steen et al. 2010), but leave out problems of sea level rise (Van Dijk 2016a). In this framework, Table 9.2 summarizes smart solutions for urban water issues.

Rainwater harvesting is promoted in Beijing, both in the centre of town and in the rural areas of the city state. However, it has turned out that at the current price of electricity, and given there is no charge for pumping up groundwater, it is cheaper to continue to use groundwater than to invest in rainwater harvesting (Liang and van Dijk 2011).[8]

9.12   Specific Flows: Waste

Cities are generating more and more waste, and in many Third World cities, the provisions to deal with the issue are not up to the standard. The recipe is reduce, reuse and recycle, using smart strategies (Binnenlands Bestuur 2007b: 20–22). However, the experience with this approach is limited in developing countries. Table 9.3 summarizes some of the possibilities to use IT in solid waste management.

Let us look at dealing with one special type of waste: e-waste (Van Dijk 2016b). E-waste is a term used to cover items of all types of electrical and electronic

---

[8] These projects are analysed below in terms of their success and governance structures.

Table 9.3 Possibilities to use IT in urban solid waste management

Issue	Smart solution	Reference
Time necessary to collect the waste	Use route planning systems	Oduar-Kwarteng and van Dijk (2007)
Are the containers full?	Put a chip in the containers	Mohammed and van Dijk (2017)
Selection of private operators	Electronic procurement makes corruption chances smaller	Experience, for example, UNIDO with this system Tilaye and van Dijk (2014)

equipment and its parts that have been discarded by the owners as waste without the intention of reuse, because this equipment has ceased to be of any value to its owners. E-waste is one of the fastest growing waste streams globally. Since the Rio Earth Summit organized by the United Nations in 1992, the concept of sustainability extends to rendering basic services such as solid waste management and dealing with e-waste.

People are afraid of e-waste because of its possible negative effects on health and because it could pollute the environment. Indicators of unsustainable service provision concerning e-waste include irregular collection, open dumping and burning of solid and e-waste in open spaces. Often collection covers a small part of the country, cost recovery is limited or not existent and one notes poor utilization of available resources with no or very limited reuse and recycling.

In many countries the informal sector is handling e-waste. The small operators undertake segregation and dismantling of collected e-waste. A survey conducted by the Asian Development Bank in 2010 showed that in a small country like Bhutan there are mainly four types of e-waste: toner/cartridges, ink, IT equipment and used batteries. Reports emphasize that most e-waste generators are not aware of the implications of improper disposal of e-waste.

Bhutan is situated between India and China. Its government sector is getting rid of its e-waste by simply auctioning it to scrap dealers. The Department of National Properties (Ministry of Finance) collects the e-waste and auctions items that the government surrenders annually. Some of it is bought by local dealers for reuse of parts, but most of it ends up in neighbouring India. What have other small countries been doing about e-waste? Let me give one example of a systematic approach to the issue.

In Singapore through voluntary efforts three companies StarHub, TES-AMM, both e-waste recyclers, and courier firm DHL have come together and have installed 200 bins across the island at institutes of education, malls, government offices, office buildings and community clubs. It is costly to mine gold, silver and precious metals from the earth and then dump them. Urban mining is a business idea whose time may not have come yet, but it will be promoted, and its time will come eventually according to the Minister for the Environment and Water Resources in Singapore.

Many municipalities in developing countries are incapable of meeting the demand for these services, including collection of e-waste and processing it, resulting in both direct and indirect negative effects on the indicators of sustainable development. The steep increase in e-waste is the result of using more computers, mobile telephones, TV sets, etc. However, the picture of how much e-waste is generated every day is usually missing. There are precious metals to recover and polluting components to collect. However, in many countries, the legislation or the implementation of policies to achieve recovery and reuse is often not there or poorly organized.

It sometimes means that the precious metals get lost because they end up in incinerators. In Bhutan developing the value chain of e-waste is difficult, given this is a small developing country. The private and household sectors often dispose of their e-waste by leaving it to the informal sector. Sometimes these informal operators do not even have to pay for it; they just collect it. Given the need for a certain minimum quantity to allow processing and given the serious investments required for this purpose, most e-waste of a small country like Bhutan ends up in India, where the value chain for the Bhutanese ends, without getting their fair share of the value. The issue for such small country is to determine the potential of reducing e-waste, reusing it or recycling it in the country itself, instead of exporting it partially illegally to the bigger neighbouring countries. Integrated urban waste management programmes need to be developed to minimize the negative effects of waste on the urban environment.

9.13   Specific Flows: Energy

Cities account for 70% of global energy consumption and contribute a similar percentage of greenhouse gas emissions (*Financial Times*, 7-6-2017). The newspaper adds that many cities promote green energy, for example, San Francisco, Frankfurt, Vancouver and San Diego, to have 100% renewable energy before 2035.

Solutions can be Smarter Grid Solutions (pioneering Active Network Management for the real-time management of grid-connected distributed energy resources). During the Copenhagen climate conference in 2009, China did not want to commit itself to reducing its CO_2 emissions; rather it suggested bringing down the quantity of energy per unit of gross domestic product (GDP). Households in China are asked through TV campaigns to reduce their electricity consumption and are encouraged to use sun boilers to heat their water (Glaeser and Kahn 2010). Table 9.4 summarizes some smart energy solutions.

In terms of the industrial sector, Beijing is now stricter in asking certain polluting and energy (or water)-intensive industries to relocate. This policy is contributing to a reduction of energy use and environmental problems in Beijing itself, but does not solve any problems at the country level. It is difficult to estimate the effects of these policies. At the climate change conference in Paris at the end of 2015, China announced stricter targets for reductions of CO_2 emissions by 2030. Although the reactions were very positive, it means the country can still go on with polluting the air in a big way for another decade!

Table 9.4   Smart solutions for urban water issues in China.

Issue	Smart solution	Reference
Generating power	Solar panels	Glaeser and Kahn (2010)
Reducing CO_2 emissions	Energy-saving performance contracts	Binnenlands Bestuur (2017b):40)
Network management and the power mix	Interoperability and the challenge of scale	
Energy need for heating or cooling	Using underground water and temperature exchange systems	
Saving on heating and cooling in buildings	Appropriate design and promoting the use of isolation material	

9.14   Other Flows: Managing Information

Cities also have to absorb flows of goods, services, people, ideas and money, and the key question is whether they can adapt to the new situation. This is called the absorptive capacity of a city, or its resilience. Townsend (2013) gives as examples:

> In Chicago, GPS sensors on snow plows feed a real-time "plow tracker" map that everyone can access. In several cities (for example Zaragoza in Spain), a card (called "citizen card" in Zaragoza) can get you on the free city-wide Wi-Fi network, and unlock a bike share. Citizens cards are also used in Zaragoza to check a book out of the library, and pay for your bus ride home. In New York, a guerrilla group of citizen-scientists installed sensors in local sewers to alert you when storm water runoff overwhelms the system, dumping waste into local waterways.

> As technology barons, entrepreneurs, mayors, and an emerging vanguard of civic hackers are trying to shape this new frontier, (his book) Smart Cities considers the motivations, aspirations, and shortcomings of them all while offering a new civics to guide our efforts as we build the future together, one click at a time.

Smart city requires infrastructure, whether the focus is on the Internet of things that may make it easier to manage our house from a distance or whether the bigger information flows at the city level are considered. The Internet of things is relatively new, but can change a lot.

Kenworthy (2006) emphasizes the role of transport in sustainable cities. He wants to move away from a car-dominated city and to build cities around foot paths, bicycle lanes and public transportation. Transport is an important flow issue, since it is crosscutting: it has to do with different means of transportation and their fuel consumption and with the design of a city. Transport can be an important source of air pollution and causes a lot of noise. In 2009, buying small cars was made more attractive in China as a way to emerge from the worldwide economic crisis, but for environmental purposes this was a step backward.

Finally, transportation is an important issue in physical planning, which can also help to deal with the issue. Integrated transport policies are desired, but often specialized departments and different levels of government deal with different modes of transportation and different types of roads. Much has been achieved in terms of reducing travel time and congestion by using information technology to track traffic and inform traffic management and travellers. Sometimes rapid transit systems have been introduced, or an alternative type of transportation has been promoted when the sensors detect congestion. Finally, the means of transportation (e.g. busses) also have tracking equipment, not just for their routes and distances covered but also for the use of energy and suggesting how to drive and resulting in more comfortable transport and less maintenance cost for the companies.

9.15   Solutions: The Smart Eco-City of the Future

How can smart eco-cities be promoted? People in Beijing have tried to make their city more environmentally friendly. The best results can be translated into climate mitigation or adaptation policies, to be implemented elsewhere. Different instruments have been used: incentives, subsidies, pilot projects and publicity campaigns. Urban development also means forging new partnerships between parties that have often not worked together before: government officials, NGOs and private sector businessmen. This requires new urban governance structures, and the urban manager may be instrumental in that. Most initiatives are started off at the level of the city, like the promotion of ecological neighbourhoods and innovative housing schemes. Others are triggered at the national level, for example, subsidies (30% of the construction cost is reimbursed in China, if certain environmentally friendly investments are made). Initiatives at the household level depend very much on the urgency of the issue and the level of awareness among the people concerned.

The complicated urban governance structures in China make it difficult to deal effectively with the consequences of climate change and to become an eco-city. The top-down government approach often conflicts with grass-roots-level initiatives. An analysis of the governance structures in place shows, on the one hand, that the implementation of policies and programmes often gets stuck at the local level. On the other hand, citizens do take initiatives to deal with the challenges that they are facing, which are partially climate change-related.

One notes a large number of disparate initiatives in Beijing to move in the direction of becoming a smart eco-city. However, the Chinese government does not stimulate enough initiatives from below. What would the ecological city of the future look like? The eco-city of the future is not just about dealing with environmental issues. Such a city will also need a sound economic basis; appropriate solutions for its transport systems, for housing and for urban service delivery; and the provision of urban amenities. There is more and more attention focused on the quality of life in cities, and the presence of sufficient urban services and amenities is an important factor in making a city attractive. A lot more is possible within existing urban systems if one is ready to think in terms of creative solutions, using experiences gained elsewhere. The challenge is to introduce environmentally and financially sustainable solutions. In China, more radical solutions are being tried, such as separating grey and brown water and rainwater harvesting.

Sustainability does not seem to be the major principle used in developing Beijing at the moment. For the Olympic Games in 2008, some efforts were made to create a greener impression of the city. However, on most of the criteria used to measure sustainable development, Beijing scores low. After cannibalizing natural resources such as land, water and clean air, Beijing is now starting to realize that a greener approach to urban development is necessary. However, simply moving polluting industries to other cities is not going to do the job. Even closing down 2000 industries, as announced in the *Financial Times* (11-8-2010), is not enough. A number of these industries will open again in the interior of the country, in the underdeveloped western part. In addition, while public campaigns on television are good, they are certainly not enough to change the attitudes of households and entrepreneurs, who have been lax as far as sustainable development is concerned.

Changes in the behaviour of consumers will be required, just like a combination of better water management, better energy management, collection and treatment of solid waste and striving towards integration. Water demand management is a good start at the household level, just like separation at source, and composting at home is a good start for ecologically friendly solid waste management. Closing the water cycle to deal with water in a more efficient way could be an option. Citizens have learned how to benefit from government programmes but also that they should not wait for government support. It is important to utilize citizens' knowledge and other resources and their capacity to deal with climate change-related issues. Furthermore, finding financing for climate change activities is important, given the increase of costs. Alternative financing mechanisms such as the clean development mechanism and tradable carbon emission right could be used. The objective is to increase the resilience of the different urban ecosystems in China. This requires better management, strong formal (public) and informal institutions, PPPs, sharing of knowledge and leadership.

9.16 Conclusions

Smart eco-cities may be a solution to the challenges of rapid urbanization, demographic change and changing energy needs. The development of smart cities should increase our capability to adapt to rapid changes, to pool knowledge, to engage in best practice sharing and to discuss initiatives taken by cities to tackle these challenges.

Smart eco-cities are focusing on managing flows, including the flow of people (migrants, poor people, farmers, etc.) and money (remittances and investments). In the smart eco-city of the future, the attention goes to managing the flows smartly, taking the interactions between different flows into account.

We learned from evaluating climate change adaptation policies Van Dijk (2013):

1. That not only output indicators but also process indicators are needed.
2. The analysis should show developments over time or through a comparative analysis.
3. There is a need for more awareness raising, particular at the household level.
4. Also waste collection and reuse/recycle require improved management at the level of the municipality or the utility concerned.
5. It is often necessary to create new governing structures, involving the people concerned, the public and the private sector.
6. Special attention is required for poor people.
7. It is important to make people conscious of these issues and the possible solutions!

Implementing smart technologies means indeed engaging with energy, water, transport, etc., data and processing it to allow smart decision-making. Smart ecological cities require integration of different information flows, and this integration could take place in the framework of urban management. However, we need smart eco-cities which start with a vision of what they want to achieve. Then the technology can be used to drive the desired change. At the same time there is the danger of creating big brother who is watching us and who would like to use the information coming out of the different flows to 'nudge' us in the desired direction. Aldous Huxley and George Orwell have written about Singapore already some time ago.

We learned from the examples that smart management requires collecting information and processing the information flows. Cities are using more and more sensors, for example, cameras or tracking equipment in busses or waste bins. In the latter case, chips are also being used to find out whether the bin needs to be emptied.

A city full of digital sensors is meant to make the city safer, healthier and more liveable. However, there is also an ethical side to big data. On the positive side, it allows activities to start without a hitch; on the other hand, it may lead to controlling citizens all day through.

If you think this is all impossible, let me quote the *Gazet van Antwerpen* (30-4-2016: 28): Smart city, one network connects everything. It reports that Singapore is working on a central computer system which would monitor everything that happens in the city. It is a high-tech system, connected to your smart phones and cameras. Already in 2014 thousands of sensors and cameras were installed monitoring every movement and activity. Integrating the information the government can spot where you dump your waste, who uses a lot of electricity or is getting together to start a demonstration against being monitored all day. In principle the information is available for everybody, but the government may want to filter who gets what. This example shows at the same time the limitations of processing all the information on all these flows in an integrated way. The question becomes who decides how the results of such integrated digital management of urban information flows will be used?

References[9]

Binnenlands Bestuur (2017a). Van uitnodigings- naar uitdagings planologie. Week 24. Journal of the VNG.

Binnenlands Bestuur. (2017b). A special on Green government. July, pp. 38–40.

de Bruijn, C. A. (1987). Monitoring a large squatter area in Dar es Salaam with sequential aerial photography. *ITC Journal, 3*, 233–238.

Florida, R. (2004). *The rise of the creative class*. New York: Basic books.

Glaeser, E. L., & Kahn, M. E. (2010). The greenness of cities: Carbon dioxide emissions and urban development. *Journal of Urban Economics, 67*, 404–418.

Hajer, M., & Dassen, T. (2014). *Smart -about- Cities, Visualising the challenge for twenty-first century urbanism*. The Hague: nai010/PBL publishers.

Hou, E. (2000). Briefing Paper on Water Governance Structure in Beijing, PRC. http://www.chs.ubc.ca/china/water%20governance.pdf

Jong, M. de. (2016). Delft: Sustainable urban and infrastructure development in China: why inter-governmental relations are the key. Inaugural address at Delft university of technology.

Kenworthy, J. R. (2006). Dimensions for sustainable city development in the Third World. *Environment Urbanization, 18*, 67–86.

Kitchin, R. (2015). Making sense of smart cities: Addressing present shortcomings. *Cambridge Journal of Regions, Economy and Society, 8*(1), 131–136. https://doi.org/10.1093/cjres/rsu027

Li, H., Gupta, J., & Van Dijk, M. P. (2013). China's drought strategies in rural areas along the Lancang. *Water Policy, 15*, 1–18. https://doi.org/10.2166/wp.2012.050

Liang, X., & van Dijk, M. P. (2009). Financial and economic feasibility of decentralized waste water reuse systems in Beijing. *Water Science and Technology, 61*(8), 1965–1974.

Liang, X., & van Dijk, M. P. (2011). Economic and financial analysis on rainwater harvesting for agricultural irrigation in the rural areas of Beijing. *Resources, Conservation and Recycling, 55*, 1100–1109.

Liang, X., & Van Dijk, M. P. (2012). Beijing, managing water for the eco city of the future. *International Journal of Water, 6*(3/4), 270–290.

Liang, X., & van Dijk, M. P. (2018). Identification of decisive factors determining the continued use of Rainwater Harvesting Systems for agriculture irrigation in Beijing. In: Rahman (Ed.) *Rainwater harvesting: quantity, quality, economics and state regulation*. Basel: MDPI. Published in 2015 in Water, an open access journal (ISSN 2073-4441), pp. 61–72.

Mohammed, A. A., & van Dijk, M. P. (2017). Practice and determinants of solid waste collection: The case of private collectors in five ethiopian cities. *International Journal of Waste Resources, 7*, 2. https://doi.org/10.4172/2252-5211.1000280

Oduro-Kwarteng, S., & van Dijk, M. P. (2007). Regulatory environment for private sector involvement in solid waste collection in Ghana. *International Journal of Environment and Waste Management, 20*(1), 35.

Paulsson, B. (1992). *Urban applications of satellite remote sensing and GIS analysis*. Washington: World Bank, Urban Management Programme.

Rotterdam. (2008). *Rotterdam, climate proof*. Rotterdam: Municipality.

Qiu, L., & van Dijk, M. P. (2014). Water pollution and environmental governance systems of the Tai and Chao Lake Basins in China in an international perspective. *International Journal of Water Resource and Protection, 7*, 830–842. https://doi.org/10.4236/jwarp.2015.710067, open access.

---

[9]You can read more detailed papers on these flows by the author and his PhD students on www.researchgate.net or www.academia.edu.

The Rockefeller Foundation and ARUP. (2014). *City resilience framework*. New York: The Rockefeller Foundation and ARUP.

Shrouf, F., & Miragliotta, G. (2015). Energy management based on Internet of Things: Practices and framework for adoption in production management. *Journal of Cleaner Production, 100*, 235–246.

Source Water Magazine. (2017). editor@thesourcemagazine.org, consulted 31-8-2017.

Tilaye, M., & van Dijk, M. P. (2014). Private sector participation in solid waste collection in Addis Ababa (Ethiopia) by involving micro-enterprises. *Waste management & Research, 32*(1), 79–87. https://doi.org/10.1177/0734242X13513826

Townsend, A. M. (2013). *Smart cities: Big data, civic hackers and the question for a new utopia*. New York, WW Norton and company, 7-10-2013.

Turkstra, J. (1998). *Urban development and geographical information, spatial and temporal patterns of urban development and land values using integrated geo-data, Villavicencio, Colombia*. Enschede: ITC Publications.

UN Habitat. (2014). *Sustainable city*. Nairobi: UN Habitat.

Van der Steen, P., et al. (2010). *Switch final report*. Delft: UNESCO-IHE.

Van Dijk, M. P., Edelenbos, J., & van Rooijen, K. (Eds.). (2017). *Urban governance in the realm of complexity*. Warwickshire: Practical Action.

Van Dijk, M. P. (2006). *Managing cities in developing countries. Cheltenham: Edward Elgar and the Chinese version*. Beijing: Renmin university press.

Van Dijk, M. P. (2009). Ecological cities, illustrated by Chinese examples. In M. A. M. Salih (Ed.), *Climate change and sustainable development, new challenges for poverty reduction* (pp. 214–233). Cheltenham: Edward Elgar.

Van Dijk, M. P. (2013). Drought policies, the experiences of China and Australia. In: Zhang J. & Voon Phin Keong (Eds.). *Climate change and sustainable development in China, policies for mitigation and reactions of farmers in the Yunan province*. Enterprise anthropology: applied research and case studies. Beijing: CASS, pp. 237–267.

Van Dijk, M. P. (2014a). Measuring eco cities, comparing European and Asian experiences: Rotterdam versus Beijing. *Asia Europe journal*. https://doi.org/10.1007/s10308-014-0405-7

Van Dijk, M. P. (2014b). Formal and informal waste collection in Chinese cities, editorial. *International Journal of Waste Resources, 4*, 3. https://doi.org/10.4172/2252-5211.1000e107

Van Dijk, M. P. (2015). Analyzing eco-cities by comparing European and Chinese experiences. In T.-C. Wong, S. S. Han, & H. Zhang (Eds.), *Population mobility, urban planning and management in China* (pp. 189–206). Berlin: Springer. https://doi.org/10.1007/978-3-319-15,257-8_11

Van Dijk, M. P. (2016a). Financing the National capital integrated coastal development (NCICD) project in Jakarta (Indonesia) with the private sector. *Journal of Coastal Zone Management, 19*(4), ISSN: 2473–ISSN: 3350. https://doi.org/10.4172/2473-3350.1000435

Van Dijk, M. P. (2016b). Can small countries benefit from the E-waste Global Value Chain? *International Journal of Environment and Waste Management, 6*(1), 1. https://doi.org/10.4172/2252-5211.1000e110

Van Dijk, M. P. (2017). Smart eco cities, managing information flows in an integrated way, the example of solid waste and waste water. *Editorial in the Journal of Waste Resources, 7*, 2. https://doi.org/10.4172/2252-5211.1000e111

Chapter 10
Reimagining Resources to Build Smart Futures: An Agritech Case Study of Aeroponics

Helen Mytton-Mills

10.1 The Problem

10.1.1 Changes in Globalised Demand

The global human population is rising at a rate,[1] and to a level, where maintaining food demand is unsustainable, and raising the nutritional standards of the poorest in the society globally is seen by many across the global food industry as a deferred challenge, to be approached as the issues present themselves, or to be managed through government and charitable aid. Many acknowledge that there is a need for proactive planning as current agriculture is already putting irreparable strain on global resources; however, the majority of the world producers are not in a position to invest to change the status quo, firstly due to the prevalence of subsistence to small producers and secondly as the government subsidies have depressed food price inflation, larger producers find themselves bearing the costs of change.

Consumer food demand in developed countries has changed from seasonal to seasonless, over recent decades, which had promoted the growth of an international integrated supply chain, increasing the food miles and consequently magnifying the environmental impact of developed countries' food consumption. This impact is compounded by increasing volumes of raw produce being processed into final products for consumption, giving rise to growing numbers of examples of where food products have greater carbon footprints than their net weight.

---

[1] UN Department of Economic and Social Affairs 2015 Report. http://www.un.org/en/development/desa/news/population/2015-report.html

H. Mytton-Mills (✉)
Aponic Ltd, Sudbury, Suffolk, UK
e-mail: helen@aponic.co.uk

© Springer International Publishing AG, part of Springer Nature 2018
M. Dastbaz et al. (eds.), *Smart Futures, Challenges of Urbanisation, and Social Sustainability*, https://doi.org/10.1007/978-3-319-74549-7_10

As well as for more year-round produce, demand has been restricted to higher quality, visually consistent produce. This has had not only an impact on the income of farmers who see restrictions on what they can sell, but it also has a negative impact on the waste inherent in the system. As the majority of livestock feed is now generated from grain crops, the market for suboptimal, although nutritionally valuable, produce or 'wonky veg' has contracted sufficiently enough to make imperfect produce less viable commercially. Food processing companies are also increasingly moving towards automation to save on rising labour costs and variable labour supply. Whereas they may previously have taken nutritionally good produce rejected by supermarket wholesalers, they now increasingly require standardised produce to feed into their machines. Growers combat these commercial pressures by oversowing, if they have enough land, to ensure that they can fulfil their contracts with wholesalers in terms of quality, consistency of size and aesthetics, and timeliness of delivery. This reduces margins, increases the pressure on the soil, and magnifies waste in the system.

Over recent decades, the increasing market dominance of supermarkets, and the resultant decline of smaller food retailers, has dictated the way that produce is purchased. Using their monopsony power, large retailers engage in buying behaviours that support an increase in monoculture and a reduction in the crop rotation and fallow system that for generations has been used to maintain even levels of nutrients in the soil. Farmers tender for sales based upon quality, and the retail wholesalers draw up requirements, with strict tolerances, punitive associated packaging costs, and the purchaser having the right to cancel the order without compensating the farmer for onward repackaging and sales costs to another retailer.[2] All of these conditions leave the risk centred on the producer rather than the retailer.

The farmer, with little to no control over the supermarket demand or the price that they will pay, is guided to grow the one, or few crops, that they have previously had a market for. This has a longer-term negative effect upon the soil, as monocropping consistently depletes a singular set of nutrients, leaving deficiencies and imbalance in the soil's biome. This forces farmers to compensate with chemical fertilisers to keep crop production viable, rather than being able to invest in the maintenance and restoration of the soil.

There are significant variables inherent in farming; the weather, which can delay or bring forward produce maturity, the prevalence of historic and new pests and disease which need to be managed at additional cost, the viability of land year on year, the cost of fuel and irrigation water, and the price variations at harvest, there is a palpable cause for the rise in smaller farming enterprises going out of business, the growth of corporate mega-farms to navigate the variability through scale, the increase in contract farmers using their machinery to fulfil the roles required on the larger farms, and the industry suicide rate being amongst the highest in all professions worldwide.[3] There has been a drive towards business diversification to mitigate

---

[2] https://www.theguardian.com/environment/2011/jul/02/british-farmers-supermarket-price-wars
[3] Farmers' suicide: Across culture P. B. Behere and M. C. Bhise. *Indian Journal of Psychiatry.* https://www.ncbi.nlm.nih.gov/pmc/articles/PMC2802368/

the fluctuating returns from small- to medium-scale farming in developed countries, but as this often is into non-food production industries, such as solar energy, hospitality, tourism, and events; this is compounding the food production demands of the future, by contracting supply rather than furthering the sustainable change required to meet the challenge.

10.1.1.1 Challenges of Resource

The main causes of challenge with maintaining food demand are the unsustainable use of resources, principally soil, water, and fossil fuels, and the endemic waste within the global system. Water is an issue of depleting groundwater supply and less predictable precipitation patterns, the impact upon soil salt levels, and in addition, the amount of fossil fuels required to clean and deliver water for irrigation.

Irrigation makes it possible to grow crops in regions where there is either not enough rainfall to meet the crops' water needs or where precipitation is uneven and crop yields can be boosted by improving access to water periodically. However, irrigated water contains dissolved salts that are left behind when water evaporates, which leads to salinisation of the soil, which can be toxic for crops at higher concentrations.

The United Nations Institute for Water, Environment and Health found that about 7.7 square miles of land in arid and semiarid parts of the world is lost to salinisation every day. In some areas, salinisation can affect half or more of irrigated farm fields. As well as crop yield losses, there are other socio-economic and environmental impacts including employment losses, increases in human and animal health problems and losses in property values of farms with degraded land, and more greenhouse gas contributions because degraded soils don't store as much atmospheric carbon dioxide.[4]

Fossil fuels are intrinsically linked to modern agriculture through seed treatments; fertilisers, herbicides, pesticides, and fungicides; the manufacture, maintenance, and running of agricultural machinery; as well as the washing, packaging, refrigeration, storage, and delivery of the produce. Unlike prevalent consumer trends towards economical and electric cars, agricultural machinery is getting bigger and uses more fuel than previous generations of tractors, drillers, sprayers, and combine harvesters. The incentive driving this direction of demand is due to the rise of corporate and contract mega-farming, which requires maximum efficiency in its operations, and the relative rise or variability in the price and availability of labour in developed countries. A reliance on fuelled, increasingly sophisticated, automated capital equipment to physically farm as well as to forecast, monitor, and predict weather events and model harvest outcomes is no longer an uncommon phenomenon amongst large agribusiness.

---

[4]United Nations University Institute for Water, Environment and Health. http://www.smithsonianmag.com/science-nature/earths-soil-getting-too-salty-crops-grow-180953163/#ezAUCl95O0MfYASH.99

Innovation in fossil fuel-centred technology is well established and has guided the progress of most industries through more efficient engines, catalytic converters to improve emissions, and better exploratory techniques to source new fields to maintain global supply. However, as OPEC operates as a cartel; it artificially manipulates world price through constricting supply (Bentley 2002, Campbell 2004), leaving agricultural producers at the mercy of any price variation. While there are concerns about the end of the world's fossil fuel supply, the industrial processes and the engineering behind the capital developments in agriculture do not treat fossil fuels as a finite resource. Although the likelihood of oil and gas running short before the payback periods on the latest steerable combine harvesters is slight, there is a definite trend towards larger-scale efficiencies rather than investment into longer-term sustainable practices. As oil and gas reserves are used, the resource becomes harder to extract and thus more expensive to obtain – this will fuel a global inflation on the price of fossil fuels.[5] This model for resource depletion is known as Hubbert's peak.[6] That depletion is expected in the short to medium term and poses a threat to food security, because with the current system, fossil fuel supply shortages mean food supply shortages.

Overall, the food industry worldwide accounts for 30% of the global energy consumption and produces over 20% of greenhouse emissions emitted globally,[7] and relies increasingly on oil to facilitate its mechanics and logistics and provide solutions for the issues that it creates; tractors, shipping, GMO seeds with fertilisers, and pest control.

We are currently in the saturation period of agriculture characterised by greater amounts of energy required to produce smaller increases in crop yield. An evergrowing amount of energy is expended just to maintain the productivity of the current system; for example, about 10% of the energy in agriculture is used just to offset the negative effects of soil erosion, and increasing amounts of pesticides must be sprayed each year as pests develop resistance to them.[8]

Phosphate is a primary component in the chemical fertiliser which is applied in modern agricultural production. However, although rock phosphate reserves will be depleted in 50–100 years, peak phosphorus will occur in about 2030 (Cordell et al. 2009). The phenomenon of peak phosphorus is expected to increase food prices as fertiliser costs increase and as rock phosphate reserves become more difficult to extract. In the long term, phosphate will therefore have to be recovered and recycled from human and animal waste in order to maintain food production.[9]

---

[5] (Bentley 2002; Campbell 2004). http://www.resilience.org/stories/2006-06-11/implications-fossil-fuel-dependence-food-system/

[6] (Gever et al. 1991).

[7] "Energy-smart" agriculture needed to escape fossil fuel trap –Food and Agriculture Organization of the United Nations. http://www.fao.org/news/story/en/item/95161/icode/

[8] (Gever et al. 1991; Pimentel and Giampletro 1994). http://www.resilience.org/stories/2006-06-11/implications-fossil-fuel-dependence-food-system/

[9] Sustainable Agriculture. https://en.wikipedia.org/wiki/Sustainable_agriculture

Waste within the system, from field to fork, is estimated to be 890,000 tonnes in the UK annually.[10] According to Action2020, 32% of all food produced in the world is lost or wasted.[11] While the resources that go into creating this produce are ultimately wasted, it is important not to lose sight, when considering change within the global system, of the opportunity for better productivity through reducing or using this waste and also the environmental impact of this waste, decomposing and releasing carbon dioxide and methane.

Soil is the lynchpin in the equation. Global produce demand increases coupled with gradual degradation of outputs has intensified the use of water and fossil fuels in agriculture[12] to compensate the sharp decline in soil nutrients.[13] The viability of potable water and fossil fuels is an issue of long-term sustainable supply; the understanding of the decline of soil is a more recent discussion, in part because the traditional agricultural practices managing the soil have given way to intensive agriculture to supply the new demand criteria of global consumption.

The debate surrounding the issue of UK soils having less than 100 harvests left within their productive capacity,[14] given the continued status quo of the current farming practices, is still young and is not fully accepted by the nation's growers, who are still experiencing a viable business model through the addition of soil treatments and irrigation. The nature of the decline in the soil organic carbon, as well as the supporting ecosystem, foretells that the production on this land will also decrease over the projected lifespan of the soil, which will put pressure on the farmers increasingly to resort to short-term measures to maintain their yields, or, if their businesses don't crumble under the pressure of commercial viability, seek new unfarmed land to cultivate, if they are not able to invest in their soil structure in a timely manner. In the UK, the ability to find available, unprotected, new, viable land is rare and expensive. Elsewhere in the world, there are established ecosystems such as wetlands, forests, and rainforests which are being eaten into at a growing rate by cultivation.

While approximately 38.6% of the ice-free land and 70% of withdrawn freshwater is already devoted to agriculture (World Water Assessment Programme (WWAP) 2009),[15] most of the unexploited land is either too steep, too wet, too dry, or too cold for agriculture (Buringh 1989). This does present farmers with little expansion options except for expanding into forest and wetland ecosystems.

---

[10] http://www.fareshare.org.uk/supply-chain-food-waste

[11] http://action2020.org/business-solutions/reducing-food-loss-and-waste

[12] Implications Of Fossil Fuel Dependence For The Food System Jay Tomczak, Tompkins Country Relocalization Project June 11, 2006. http://www.resilience.org/stories/2006-06-11/implications-fossil-fuel-dependence-food-system/

[13] Professor Nigel Dummett, University of Sheffield. http://www.independent.co.uk/news/uk/home-news/britain-facing-agricultural-crisis-as-scientists-warn-there-are-only-100-harvests-left-in-our-farm-9806353.html

[14] Urban cultivation in allotments maintains soil qualities adversely affected by conventional agriculture. Authors Jill L. Edmondson, Zoe G. Davies, Kevin J. Gaston, Jonathan R. Leake. *Journal of Applied Ecology.*

[15] Sacks W. J. *Crop Calendar Dataset. Center for Sustainability and the Global Environment.* University of Wisconsin-Madison. http://www.sage.wisc.edu/download/sacks/crop_calendar.html

Agriculture is estimated to be the direct driver for around 80% of deforestation worldwide.[16] Slash and burn practices of land clearing for shifting cultivation are particularly damaging, as not only is the carbon stored in those forests being released into the atmosphere, but the topsoil structure held together by the roots is often so thin that when exposed to farming it can be washed away under the usual climactic conditions of the area.[17] This process has intensified since the 1970s but small-scale slash-and-char, within clearings rather than on the edges of forests, has been part of the Amazonian cultivation practices for at least 6000 years, which with the addition of organic matter produces terra preta, one of the richest soils on Earth and the only one that regenerates itself.[18]

While the destruction of habitat is of grave ecological and environmental detriment, the loss of the topsoil soon after the land is prepared for livestock rearing, or cultivation, reduces the ability for vegetative growth to prosper. In addition, the silt run-off clogs and damages waterways and further upsets another local ecosystem as well as disrupting agricultural and hydroelectric water supplies.[19] These two factors, when combined, mean that the land quickly loses its productive capacity and the farmers feel compelled to repeat the cycle and prepare more land for agriculture. For the land now left behind, the absence of topsoil leads to desertification of the ecosystem, cementing the areas' inability to regenerate its ecosystem despite no further intervention by man.

10.1.2   Effect upon the Environment and Biodiversity

The consequences for ecosystems, biodiversity, and the environment of changing nothing in terms of agricultural practices are startling in their proximity and the likelihood of being irreversible, across widespread areas of the globe. Agricultural areas in a temperate climate, such as Norfolk in the UK, are already experiencing depleting groundwater supply. Throughout the Cambridgeshire Fens, an area characterised by an abundance of water, agricultural irrigation can be seen throughout summer. In Western USA, the Colorado River has had so much water diverted from it that it no longer reaches the ocean, and the great Ogallala Aquifer is being overdrawn at 130–160% its recharge rate (Pimentel and Giampletro 1994). Approximately 90% of US agricultural lands are losing topsoil above sustainable rates (1 t/ha/yr) due to erosion, and the application of synthetic fertilisers actively promotes soil degradation.[20]

[16] Agriculture is the direct driver for worldwide deforestation Wageningen University and Research Centre. https://www.sciencedaily.com/releases/2012/09/120925091608.htm

[17] Deforestation: Facts, Causes and Effects Alina Bradford Live Science. https://www.livescience.com/27692-deforestation.html

[18] Sustainable Agriculture. https://en.wikipedia.org/wiki/Sustainable_agriculture

[19] Soil Erosion and its effects Rhett Butler. https://rainforests.mongabay.com/0903.htm

[20] (Gever et al. 1991; Pimentel and Giampletro 1994). http://www.resilience.org/stories/2006-06-11/implications-fossil-fuel-dependence-food-system/

When fresh water supplies are depleted and irrigation changes the salt and nutrient profile of a soil, it has a knock-on effect on the delicate biome of that ecosystem. The full range and function of flora, fauna, fungi, and microorganisms that inhabit our soils are not yet fully understood. This creates an uncertainty over how the treatments and processes that we apply to soil, through agriculture, will affect the health and productivity of this finite resource. Extinction will not just be an issue of the rainforests, but we may lose the microorganisms that make up the biome globally, with unknown consequences, from not yet understanding the role that they played in the environment where they existed. Other problems stem from the vast amount of pollution associated with agricultural run-off, which degrade aquatic ecosystems and create dead zones in the ocean (Matthews and Hammond 1999).

Crops only absorb about half of the nitrogen they are exposed to, much of the rest runs off the fields with water flow, saturating the environment and polluting aquatic ecosystems (Matthews and Hammond 1999).

The challenges and the localised decline of bees and other pollinators globally in association with insecticides are a debate, as well as a phenomenon, that has pitted the farmers against the policymakers, in the absence of other available crop security measures. The rise of mega-farms of grain production rather than insect pollinated food crops, raises nutritional issues as grains and starches are solid dietary staples but provide little of the vitamins and minerals that humans require to survive. They are also a poorer quality of dietary fibre, and the more processed and refined the flour, the less positive effects upon the digestive bacteria, insulin production, and detrimental cholesterol.[21] Over the recent decades, there has been a seismic increase in people suffering intolerances to grains, specifically those containing gluten, in line with the wide-scale industry use of hybridised grain crops designed to improve crop success rates. It is possible that these strains contain a different type of gluten not suiting a set of digestive enzymes based upon a long-term gastric evolution based upon heritage grain varieties.[22] If this is a trend set to continue, then the decline of pollinators and pollinated produce is of real concern in how to source food, outside of meat and dairy, which leads to an incomplete and highly unsustainable diet.

While this chapter focuses entirely on arable agricultural produce, it is important to note that the practices of pastoral farming for meat are well publicised as a highly unsustainable food source. Chickens are now the most common bird in the world, entirely due to farming, and as much of the agricultural deforestation globally is for beef pasture as it is for arable cultivation.

All of the current agricultural practices, from both the demand and the supply side, in order to sustainably feed the world's growing population, and methods for growing food have to evolve.[23] Amongst a number of the important global changes

---

[21] http://blog.aicr.org/2015/01/27/study-effect-of-whole-grain-oats-on-gut-bacteria-and-health/

[22] Did Wheat Hybridization Give Rise To Celiac Disease? http://www.einkorn.com/did-wheat-hybridization-lead-to-celiac-disease/

[23] Comparison of Land, Water, and Energy Requirements of Lettuce Grown Using Hydroponic vs. Conventional Agricultural Methods. https://www.ncbi.nlm.nih.gov/pmc/articles/PMC4483736/

required, economic and social well-being must improve for that large fraction of the world's peoples now in poverty. This includes more and better food. A doubling of the population would necessitate the equivalent of a tripling, or more, of our current food supply to ensure that the undernourished were no longer at risk and to bring population growth stabilisation within reach in humane ways, without widespread hunger and deprivation. Improved nutrition may be achieved by dietary shifts and improved distribution as well as by an increased quantity of food.[24]

10.2 Mainstream Approaches to Improving the Sustainability of Agriculture

The issue of sustainability has not long been top of the agenda internationally. Today's farmers worldwide are inheriting the legacy of the Green Revolution, which was designed with increased outputs to feed a growing world population at a value that could sustain commercial agribusiness. It achieved this by using large amounts of fossil fuel energy in the form of synthetic nitrogen fertilisers, petroleum based agrochemicals, diesel powered machinery, refrigeration, irrigation, and an oil dependent distribution system (Gever et al (1991)). This system destroyed biodiversity, contributes to global climate change, and degrades soil and water quality.[25] Perversely, historic farming practices were more focused towards soil and resource husbandry, and a number of the approaches to safeguarding the future of agriculture hark back to traditional methods.

10.2.1 Crop Rotation

Crop rotation involves changing through a predetermined pattern of crops, in successive years, on cultivated land to balance the nutrients within the soil over time, rather than one crop type leeching a singular set of nutrients, and to reduce the prevalence of pests, who tend to attack specific crop types.

10.2.2 Soil Amendment

Recycling from food production and domestic waste can be used to produce low-cost organic compost for organic farming. There is great progress in the field of Black Soldier Fly rearing to process organic waste into soil additives with improved nutrient bioavailability for plant uptake.

---

[24] Constraints on the Expansion of the Global Food Supply by Henery W. Kindall and David Pimentel, from Ambio Vol. 23 No. 3, May 1994, The Royal Swedish Academy of Sciences.

[25] Implications of Fossil Fuel Dependence for the Food System. Jay Tomczak, originally published by Tompkins Country Relocalization Project June 11, 2006. http://www.resilience.org/stories/2006-06-11/implications-fossil-fuel-dependence-food-system/

10.2.3   Perennial Crop Cultivation

Growing a diverse number of perennial crops in a single field, each of which would grow in separate season so as not to compete with each other for natural resources. This system would result in increased resistance to diseases and decreased effects of erosion and loss of nutrients in soil (Glover, J. D., & Reganold, J. P. 2010). Crop management would need to be more highly skilled in order to create commercially productive outputs from this system.

10.2.4   Perennial Grain System

Today, most of humanity's food comes directly or indirectly, as animal feed, from cereal grains, legumes, and oilseed crops, all of which are annual crops. Replacing some of the single-season crops with perennials would create large root systems capable of preserving the soil and would allow cultivation in areas currently considered marginal.[26] Perennial plants reduce erosion risks, sequester more carbon, and require less fuel, fertiliser, and pesticides to grow than their annual counterparts (Cox et al., 2006; Glover et al., 2007).

In recent years, plant breeders in the USA, Argentina, Australia, China, India, and Sweden have initiated plant genetic research and breeding programmes to develop wheat, rice, corn, sorghum, sunflower, intermediate wheatgrass, and other species as perennial grain crops (Glover and Reganold 2010). However, it could take 20 years to develop perennial wheat ready to be widely planted on farms. At present, it takes plant breeders more than a decade just to develop new varieties of annual wheat and ensure that they are ready to be widely grown for commercial use (Glover and Reganold 2010).

10.2.5   Government Subsidies

Government legislation forms an important incentive for changing farming practices. The creation and maintenance of nature strips as part of the CAP subsidy provide a permanent investment to ecology, but also permanently reduce the acreage of agricultural land, which impedes the long-term ability to escalate food production. Also, while biodiversity is able to regenerate and coexist with the field environment, they are still affected by run-off nitrate and overspray.

Payment for ecosystem (or environmental) services (PES) schemes are a novel idea mooted for consideration. These schemes reward the farmers financially for being stewards of their environment, conserving biodiversity and supporting the

---

[26] (Cox et al. 2006; Glover et al. 2007).

calculated 'value' of their local ecosystem in terms of natural, socio economic, community, climate change, and water management benefits.

However, paying farmers to be more environmentally friendly won't solve the problem of food security, and if these schemes reduce crop yields, it may result in increased production elsewhere, displacing the impacts that subsidies are paying some farmers to mitigate (Tanentzap et al. 2015). It is expected that at a worst case scenario, this results in further land being sucked into the agricultural churn.[27]

10.2.6 Selling Locally

One of the biggest causes of greenhouse gases is transportation, and growing food for consumption in other parts of the country, or even internationally, requires a lot of fossil fuels. Selling food in local markets helps reduce these emissions. There are other benefits as well: food sold locally needs less packaging, as it does not have as long between harvest and sale as food grown for distant markets. Selling locally also keeps money in the local economy and allows producers to engage directly with their customers. This fosters good community relationships, with customers more likely to support local producers, meaning even those farming on a small scale can make a living.[28] However, in order to engender change beyond the novelty of local, consumer choices need re-educating with regard to seasonality.

10.2.7 Growing Organically

The organic approach uses biological processes to achieve high soil quality, control pests, and provide favourable growing environments for productive crops, by the prohibition of use of most synthetically produced inputs. For farm products to meet organic standards, farmers must use 'organic' inputs, which are usually expensive. Organic farmers commonly use cover crops, legumes, compost, animal and green manures, and animal by-products (fish, bone, and blood meals) in their soil-building and nutrient management programmes (Vandermeer 1995).

Organic farms often have smaller nutrient surpluses than conventional farms.[29] The lower leachable nitrates in organic systems could be because they operate at lower levels of nitrogen application and because nitrogen in organic systems is bound to organic fertilisers, such as composts and manures, when added or incorporated in the soil. (Kasperczyk and Knickel, 2006; Kustermann et al., 2010).

---

[27] http://www.cam.ac.uk/research/news/paying-farmers-to-help-the-environment-works-but-perverse-subsidies-must-be-balanced

[28] https://www.regenerative.com/magazine/six-sustainable-agricultural-practices

[29] (Kasperczyk and Knickel 2006; Kustermann et al. 2010).

Organically managed soils have been shown to store nitrogen more efficiently than their conventional counterparts.[30]

The drawbacks of organic as a business model include weed control, as one of the greatest challenges to yield productivity and economic profitability in organic systems. Although consumers often perceive organic fruits and vegetables as more nutritious than their conventional counterparts, the nutritional superiority of organic crops has not been unequivocally demonstrated. All of these issues decrease the commercial viability for small- to medium-scale farmers to take up growing organically, especially if certification is not able to be maintained, except for the higher prices that organic certification can bestow upon the produce. Unfortunately, if greater take up of organic practices were to happen, an increase in supply would depress the higher prices that make growing organic attractive as a business model.

10.2.8   Cover Crops

Increasingly in vogue, cover crops form an important function in successful rotational practices. These consist of plants, sown not for harvest, but to convert solar energy into the ground, to store up nutrients and carbon in the soil and fix nitrogen levels, if the cover crop is a leguminous. They have the added benefits of opening up soil structure where compacted, and binding loose soil structures which can help prevent erosion. Better soil structures, with improved biomass, manage to absorb precipitation better, reducing run-off and improving the moisture capacity within soil. Legumes, as cover crops, are particularly useful for shoring up nitrogen within the soil, meaning that less nitrogen-based fertilisers are required in following crop cultivation. However, legumes do not overwinter well, and the farmer is faced with an opportunity cost of a productive growing time for a crop, by investing in this practice. This is true for all cover crop practices, although traditionally winter crops are less commercially valuable than the summer ripening produce and grains. While ground cover will reduce the incidence of other plants, management will still be required to mitigate weeds that establish and lay seeds in the ground which may compete with future crops. Another issue for cover crops is that sowing dates are quite specific – poor establishment often results from late sowing or low soil moisture.[31]

Agriculture and Horticulture Development Board (AHDB) funded research has shown that cover crop practices improve worm numbers and microbial activity. In the longer term, regular use of cover crops can raise soil organic matter content. Work undertaken within the NIAB TAG New Farming Systems programme has also

---

[30](Clark et al. 1998). https://www.nap.edu/read/12832/chapter/8#227

[31] https://cereals.ahdb.org.uk/media/655816/is41-opportunities-for-cover-crops-in-conventional-arable-rotations.pdf

demonstrated rotational yield and margin (over nitrogen) improvements from the use of specific cover crop approaches.[33]

10.2.9 Replacing or Balancing the Use of Fossil Fuels as an Energy Source

Dedicated fuel crops such as *Miscanthus* and switchgrass, can be grown on lands that might not be suitable for other crops and can provide an additional income source for farmers, but they are a single-market commodity. If they are to be used for ethanol production, their demand will depend on oil price and the percentage of ethanol that can be blended in fuel.[32]

This practice is good for perpetuating the fuel-based relationship with agriculture and mitigating the inflationary effects of declining oil supplies, but in practice takes land away from food production, which could be achieved through soilless methods.

10.2.10 Ways to Increase Sustainable Approaches[33]

Ensuring immediate benefits: While environmental soundness and resilience are paramount, farmers must experience an immediate benefit if they are going to change their practice to long-term sustainability. Getting benefits from sustainable agriculture is not always quick though, as it takes time for new approaches to be adapted to different agroecological and socio-economic conditions and to show their impacts: rebuilding organic matter dramatically improves soil fertility and moisture, but it can take 2 or more years for this to happen.

Providing intermediate, appropriate technology: In order to be attractive, sustainable practices need to be technically as well as economically efficient. Intermediate technological solutions such as light machinery and affordable tools can encourage small-scale farmers to test them.

Carrying out research and providing technical assistance: Farmers know a lot, but they may not know about alternative options if they have not been introduced to them. They need more training on 'nonconventional' farming methods and on innovative ways to share their knowledge while ensuring that farmers' interests and learning skills are prioritised.

Increased policy support and leadership: Addressing technical and financial constraints is important, but policy coherence is essential for scaling up. One way to reinforce policy advocacy for sustainable agriculture is by producing and consoli-

---

[32] https://www.nap.edu/read/12832/chapter/8#252
[33] Growing sustainable agriculture in Mozambique (April 2015), Laura Silici and Lila Buckley, IIED Backgrounder. https://www.iied.org/five-ways-make-farming-more-sustainable

dating evidence of its benefits, in contrast with the negative impacts of high-input intensive monocultures. A better shared understanding of these issues would provide common ground for local actors to pursue the changes that are needed in agricultural policy and practice.

The main source of the problem of take up of sustainable agricultural practices is that the right conditions don't exist for the vast majority of growers. The incentives for sustainability are simply not in place.

10.3   Soilless Growing

Growing without soil seems a contradiction to the very nature of growing. Nature uses soil abundantly, and plants have developed a harmonious symbiosis/symbiotic harmony with the different soil structures and compositions available worldwide. However, nature very rarely cultivates soil and as such operates close to a purist permaculture model of growth, where the objective is long-term species survival, rather than productive harvest yields.

When we look at the function of soil in agriculture, it has a few distinct functions and a selection of drawbacks. It provides a growing environment for the plant, giving physical stability, a medium to store water and some nutrients, as well as giving the roots a relatively stable temperature, or at the very least not exposing them to quick adjustments in diurnal temperatures.

The drawbacks of soil are much more complex. They provide an environment for other flora and fauna, from bacteria and fungi, which may support plant function as well as blight it, to plant predators and other plants which compete for light, water, and nutrients. The soil composition limits what crops can be grown successfully due to nutrient profile, water retention and drainage, and its ability to maintain structural and nutritional stability over successive harvests.

Some of the methods that are associated with soilless growing have their roots in antiquity and mythology. Famously, the Hanging Gardens of Babylon were reputed, perhaps erroneously, to be hydroponic. However, they are more likely to have been an early forerunner of a structured irrigation model that led into the development of a truly hydroponic growing environment. In truth, nature has provided the closest example of soilless growing, with the success of aquatic plants, as well as tree-dwelling orchids.

10.3.1   Aquaponics

Aquaponics is a system that best encapsulates the carbon cycle within nature. Waste from fish, and other aquatic life, is broken down within adequately oxygenated water. This oxygenating breakdown takes nitrites from the waste and transforms it to nitrates, which plants can take up as a nutrient for growth. Fish waste also has other trace elements, which make up a fairly balanced nutrient profile for the plants, and although they don't produce/synthesise any iron, the enzymes within their digestive systems break down these chemical compounds to be more bioavailable, allowing the plants to absorb them more readily. The removal of the waste from the water is beneficial to the fish, reducing toxicity and improving quality and longevity of life.

For fish husbandry as a productive output, the ability to grow plant crops from the waste products of the process is a benefit to farmers, who have a diversified crop output and a nutritional balance between protein and vegetative produce. Commercially, there are very few applications where it is possible to use fish as part of a vegetative production process, as the fish husbandry needs to remain the focus of the process rather than the needs of growing a plant crop with superlative, and consistent quality.

10.3.2 Hydroponics

Hydroponics moves away from the idealistic notions of capturing the full benefits of the carbon cycle, by replacing fish with a soluble, complete nutrient, which can be targeted specifically to the optimal needs of the plants being grown.

Set ups for hydroponics are not new to the market: there was a great movement in the 1970's for the Dutch bucket system, but this did not translate the ambition to use closed-loop systems commercially. There are plenty of modern examples of how the technology has improved, with proven results in water saving, plant growth efficiency improvements, planting density gains, and reductions in nutrient inputs.

There are drawbacks to hydroponics however. All of the systems require continuous electrical power to operate, primarily to ensure that the liquid medium for growth remains adequately oxygenated, which is an additional cost to the system over soil. One of the greater engineering challenges for these systems is ensuring even oxygenation throughout the growing medium. Typical examples of weaknesses within the hydroponic systems are where anaerobic 'dead spots' occur and plants become weakened by nitrifying bacteria and are thus more susceptible to disease, as well as growing suboptimally.[34]

The expense of monitoring, maintaining, or replacing the medium also adds an additional cost, although the cost in reducing fertiliser and pesticides, as well as eliminating herbicides, mitigates the additional costs. There is also a requirement for education and knowledge transfer to the operator; although as the majority of the tasks automated, there is additional time to take on the necessary water chemistry and engineering skills.

---

[34] http://www.osmobot.com/blog/the-importance-of-dissolved-oxygen-in-hydroponics

Hydroponics has been used commercially in strawberry growing and in Holland for produce over a number of years. Only recently have closed-loop systems entered into the market after the costs and availability of water have become a more widespread global issue.[35]

10.3.3   Aeroponics

Aeroponics differs from aquaponics and hydroponics in that it doesn't use a solid or a fluid medium to grow the plants in. The plants are supported, with their roots suspended within an air chamber, which is then sprayed or misted with nutrient-laden water. The spraying can be done at intervals, rather than continuously, as the air acts in the same manner as freely draining soil, by allowing oxygenation to the roots and a space for excess water to run off to. As there is no medium to compact, the incidence of plant disease is decreased, and stronger plants are better able to deal with any foliar pests that might be encountered in a field location.

Aeroponics is arguably the most sustainable of all of the soilless growing methods, as it uses the least energy by virtue of moving less water and not needing to oxygenate the medium and requires no replacement or maintenance to the growing medium. Both of these benefits not only reduce the cost base of operating this growing method but also reduce the carbon footprint of the final product.

Aeroponics can be adapted to a nutrient delivery system that runs either on an aquatic basis or with a hydroponic nutrient solution. This is by virtue of the growing chamber being discrete from the liquid nutrient reservoir. A defining benefit to aeroponics is that the chambers can be vertical rather than horizontal, turning acreage into volume, and utilising gravity for draining away excess water, rather than pumping water out of a horizontal system.

The design flexibility of creating modules for nutrients separately to those that contain the plants allows companies to offer a commercial-scale farming system using aeroponics, which do not fall into the traps of scaling aquaponics and hydroponics.

By using petrochemicals more efficiently through targeted nutrient usage and plastics for the products, it is possible to make more long-term use of fossil fuels by locking up the fuel for successive harvests and freeing up the finite supply to fulfil other industries' needs.

---

[35] https://www.ncbi.nlm.nih.gov/pmc/articles/PMC4483736/. Comparison of Land, Water, and Energy Requirements of Lettuce Grown Using Hydroponic vs. Conventional Agricultural Methods.

10.4 Impact of Applications of Aeroponic Growing

10.4.1 Applications of Aeroponics to Rural Economies

Aeroponics is by definition a light weight and flexible growing method, as it does away with the sheer weight of traditional growing methods and allows localised resources to act as the main fuel to run the growing systems. This enables rural economies to be self-sustaining in terms of using rainwater as the water source and solar power to run a low-power system. The modular nature of aeroponic systems allows people to start farming at any level and size, but achieve the same results as commerical scale growers, as the method is fixed and repeatable.

Three socio-economic areas will be considered with the same methods applied throughout all of them.

10.4.2 Developed Countries

In a scenario where aeroponic farming is used in developed countries, it is viewed presently as a supplemental income to farmers to help spread the annual lottery of traditional arable and livestock farming with a reliable, dependable yield and consistent quality even out of season. Many more farmers are looking for a low input, high value output method of growing food for their existing markets, or a more local market to diversify their farm earnings, especially with the advent of changing climatic conditions making traditional farming less predictable. Although much of this activity does not attract subsidies, there are many rural development grants available for diversifying farmers which allow for capital outlay to be covered, which is making diversification more attractive. The take up of aeroponic practices is often led by younger members of the farming family, as they have grown up in a culture of fast changing technology being accepted. Their starting systems are compatible with existing farm operations as it takes up comparatively little room and is highly productive. As the venture grows, by virtue of a comparably short return on investment period, it is able to employ more people in rural communities in production and selling or delivering the produce. Local markets often include hotels, restaurants, local shops, and specialist local producers. Hotels and restaurants have to travel many miles a week to buy fresh produce from centralised, larger city markets to ensure top quality food for their customers. They are enormously enthusiastic at the prospect of being able to source fresh produce that is grit free, nutritionally complete, with good flavour, and has travelled many less miles than wholesale food and so arrives fresher and with the prospect of a much longer shelf life.

Developed countries have a higher demand for exotic food and out-of-season food. When growing under cover, this can be catered for with little science being applied, to match the nutrient demands of the crop and supplemental lighting, so that the food can be produced locally to a very high standard. This attracts people with a farming or horticultural background to take advantage of higher out-of-season food prices. Urban farming is a viable option for aeroponic growers; its lightness allows for roof space to be utilised for growing where excess heat and carbon dioxide are often lost to the atmosphere. Large, unused buildings can be used for undercover growing with a ready market in towns for the produce all year round. The low capital outlay to start growing means that people with growing knowledge or experience can successfully launch a farming business without having to invest in land or machinery to supply local or niche markets.

10.4.3   Developing Countries

This is where easily attainable, low-tech growing methods work for entrepreneurial small-holder farmers and allow them to add real value to the food chain in blossoming communities and towns. Very often electricity, fuel and other common resources are unreliable and expensive; therefore a growing method that uses low-power electrical components which can be powered by solar power and trickle fed batteries with very little effort can be utilised to good effect. These farms can be run by emergent farmers or local cooperative groups which can attract grants and subsidies to grow staple products that benefit their country.

These farmers typically farm on poorer soil with irregular or contaminated water supplies. With the advent of LED lighting, UV sterilisation is attainable on the same 12-volt system that can run the aeroponic equipment which enables rainwater to be collected rather than local water courses being used to supply the irrigation water. An aeroponic system typically uses 90% less water than traditional farming, as it recirculates the water in a closed system. This means that the nutrient is also captive and gets used thoroughly, rather than washed off into the local water table. The water supply can be UV sterilised before entering the system; increasing hygiene levels and ensuring constant nutrient conditions which benefit the growing plants.

Although in developing countries the middle classes tend to be the prime movers in the process of setting up farms, the low outlay and low space requirement of vertical aeroponics mean that less affluent farmers can compete with direct parity in the market, and with the same quality of food as the larger farms.

10.4.4   Deprived Countries

In deprived countries where wars, famine, natural disasters, and adverse climate shift have affected people's ability to sustain an adequate diet, aeroponics can represent a lifeline that turns challenging circumstances into manageable scenarios once sustainable food production is introduced. Charities and NGOs collect large sums of money, but once they have installed running water, their efforts can stall as there is little infrastructure to pin further aid and growth to.

The donation of simple aeroponic growing systems allows families to grow enough food to feed their families and even start to produce a small surplus to sell locally. These systems are low-tech, and as the procedure is very repeatable, simple instructions can be followed to consistently grow food with good nutritional value without having to turn the soil, battle with soil-based pathogens or fertilise, crop rotate, and sacrifice much of the available water supplies that are needed for drinking, cooking, and sanitation. One system can produce large quantities of food and run totally off-grid. Once food is being sustainably produced, other aid efforts can concentrate on building infrastructure rather than just delivering consumable food.

Agriculture plays a large part in the culture of Zambia, where there is still 60% poverty. The local people, although able to cope with life generally, are perilously close to disaster with every harvest. They predominantly grow maize: tending not to fertilise or rotate crops which give rise to so called hidden hunger; where they are eating food with very low nutritional value. With the introduction of vertical aeroponic growing equipment, they are able to grow food with a high BRICs score and are eating high-quality food. This also means they are producing clean, consistent food to eat and sell to tourism venues such as hotels and lodges at export quality food, which can expand their horizons commercially.[36]

10.4.5   Aeroponics as a Means of Increasing GDP

A vertical aeroponic system has been introduced to Jamaica, an island experiencing social unrest and low incomes on one side of the island. Holiday hotels and beaches that cater for affluent, western holidaymakers, largely unaware of the dichotomy of fortunes, kept out of sight. The clean water on the island is largely from inland waterways. Forty-six percent of rural housing has water connection, and even less have basic effluent removal (Caribbean Regional Fund for Waste Water Management 2015).

Agriculture produces 7% of the GDP but is mainly restricted to sugar cane and bananas although rice and other staples are grown there too. The holiday trade demands high-quality, European and American-style foods, which are largely imported. This means that the 2 million annual tourists are being largely fed with imported food. The Jamaican government is actively sponsoring local community groups to grow high-quality food products to suit their visitor's tastes which will help to absorb some of the unemployment in the less affluent areas of the island and increase the country's GDP to boot. This is being made possible by the setting up of aeroponic systems in open, but netted areas, where western fruit, vegetables, salad crops, and herbs can be grown all year round, without treating the soil to adapt the

---

[36] Jason Hawkins-Row, CEO Aponic Ltd. – Observations of agricultural opportunities in Zambia 2017.

nutritional profile to match the western crops that are desired. Outside of the tourist season, these can be turned over to growing local crops or export crops for Haiti and mainland America.

10.4.6 Inclusive Farming and Social Mobility

Aeroponics removes the need to turn the soil, and it also removes the requirement for weed control and makes planting and harvesting a simple operation carried out by standing up on a solid surface usually in clean conditions. Operating staff are far easier to find and employ, and it also opens up the doors of employment to the physically impaired and people with learning disabilities to carry out socially and economically meaningful tasks for a wage.

The vertical aeroponic systems can be mounted at any height to allow the operators to work comfortably and easily through a crop with wheelchairs or other walking aids, and carry out most tasks without modification or additional support. This results in valuable capacity additions in the workforce and can create long-term opportunities for development within the business.

This also means that in countries where traditionally women, children, and elders are given more menial tasks around the farm or smallholding, they can be better utilised in tending and planting rather than constantly raising water, weeding, or tending the crops.

Aeroponic growing can enable people to produce and provide for their family, and have surplus food for selling at local markets and local outlets, as well as create a business with what they have at hand, which can sustain and enrich them. It also improves access to education and entrepreneurial activities, through a reduction in the time spent on creating a subsistence level of nutrition.

10.4.7 Application of Aeroponics for Competition Amongst Food Producers

Commercial food producers are constantly looking for opportunities to reduce overheads, logistics, and indirect costs while offering a better quality product. This has taken the form of globalised food production and supply, so that food is grown in countries that can provide out-of-season growing characteristics. The logistics that marry up international weather patterns, to predict output for peak buying seasons are highly complex and add additional cost into the farming model in order to be able to generate gains from market and environmental predictions.

Technologies like aeroponic farming have started to turn commercial heads towards efficient growing that is predictable and repeatable and that can be sited near the food supply hubs. The process can be mostly automated so that the planting by pick-and-place machines, automatic environmental control, harvesting, and packing is no longer a thing of science fiction, and becomes cheaper and easier every year.

This has caught the attention of the supermarket buyers. Through electronic portals, they can see, at any time, the readiness of the crops that they have ordered, trace the crop back to a specific seed batch, and monitor any crop protection used in the crop. They can place an order, have it delivered to their chosen hub, and have confidence that it will be on time, with the correct size and shape, and cleanliness. With growers operating in a fully managed environment, crop maturity can be sped up or slowed down in direct response to consumer demand, related via the supermarket. This not only reduces consumer waste but ensures that the grower does not supply products in tandem with other producers which would decrease the sale price through oversupply.

For the growers, this means that they can grow to order and they can plant without huge contingencies, as they are more in control of all of the variables of climate, heat, light, nutrient values, and atmospheric supplements like carbon dioxide, which can speed up growth or slow it down predictably.

With direct contact from farmer to buyer in electronic format, the supply chain can become efficient, manageable, and profitable for both farmer and buyer. The farmer can monitor and adjust his growing parameters remotely and can aggregate that same data to allow buyers to order or buy the produce to exact time schedules.

These factors mean that people can start producing food commercially without having to inherit land. Traditionally, getting started in farming meant having to buy, or take over, land and machinery and employ labour. All of which eliminate most people from the prospect of farming, as return on investment is likely to take up to a generation to materialise. Aeroponics allows entrepreneurial people to create food in industrial units, on contaminated land, or in unused barns, in such a way that they can start small and expand to any size without penalty by building on what they have continuously, with a short return to breakeven and profitability.

10.4.8 Diversification through Changing the Business Model for Existing Growers

Aeroponic growing removes many of the variables from farming to climate, soil quality, weed incursion, and insect damage and replaces them with clean, predictable crops which can be produced 365 days a year to exactly the same standard.

Under cover growing, in order to extend growing seasons, gives rise to multi-use buildings such as lambing sheds, which are only used for 2 months of the year traditionally. It can also utilise industrial units, grain stores, storage barns, rooftops, and set aside land, as well as brownfield sites and even fields that have solar panels, by growing on frames attached to the back edge of the panels.

This flexibility increases land use and efficiency throughout the growing industry and makes unusable growing land highly productive.

10.4.9   Environmental Impacts of Applications of Aeroponics

Aeroponics saves 90% of the water of traditional growing, cuts fertiliser usage by 30–50%, and yet grows faster and larger yields because of the optimum balance of fertiliser, correct pH values. However, the real secret of aeroponics is the exposure of the roots to a full supply of oxygen which allows the plant to efficiently form its oils and sugars properly. In practice, this method has no nutrient run-off to negatively interrupt the natural biome and balance of nature. When grown in biosecure greenhouses or buildings, any crop protection needed can be targeted directly at the problem rather than applied with broad sweeps which often results in damage to the natural balance of flora and fauna.

Although aeroponics is not an answer to broadacre crop growing, it can certainly relieve the pressure on soil so that it can be used less intensively and be allowed to naturally recover its organic content. Aeroponic equipment can even be placed over a field with a cover crops growing in it, and the field will still be productive while it recovers its nutrients.

For growing raw materials like cotton, it can very easily be grown efficiently in enclosed areas. As a crop, cotton accounts for 7% of the world's commercial crops and 40% of the world's pesticide usage. This high level of pesticide usage has a knock-on effect of disturbing the balance of pollinating insects and decreasing biodiversity in the areas where it is grown.

Strawberries are a good example of a crop that needs a very specific balance of nutrients and is prone to fungal diseases. Vertically grown strawberries give great yields in aeroponics and can be given a low-nitrogen, high-calcium feed, which produces the very best fruit, and in vertical grow tubes, water does not gather in the crown of the plant, and crown rot is significantly reduced. The need for fungicidal treatment to be used on the plants is kinder to the surrounding wildlife and is a high value selling point on the supermarket shelves.

10.4.10   Research Possibilities for Aeroponics

Aeroponic growing is a very suitable method for the development of desirable traits in GM crops as the crop is grown faster, so more generations can be observed in less time. Unlike crops grown in soil, the feed regime can be exactly formulated and consistent yet is repeatable and reliable so that modified crops can be grown in the same system as control crops with directly comparable results.

This gives accurate and meaningful laboratory results to study, by removing the variables thrown up in soil. The benefits of aeroponic growing echoes the aims of genetically modified crops, which are usually to reduce water usage, increase disease resistance, and increase yield. There is an additional environmental advantages to using aeroponics with GM crops, in that they are completely removed from the natural environment, and therefore concerns for their effect upon the localised ecosystems of farms can be mitigated against. With inherent improvements in yield and plant health, research can continue to investigate nutritional and flavour improvements to plant strains and varieties.

10.5   Conclusion

In order to maintain or improve nutritional standards globally, there need to be sustainable options in order to avoid the continued sacrificing of global resources, complicit with long term environmental and ecosystem damage and destruction, for short term maintenance or capacity gains within the outputs of the global food industry. Future proofing a sustainable global food supply is a long term commitment to change, with flexibility built in to adapt to unknown climactic and commercial challenges. However, growers across the world need to continue to subsist or make a living, so instant technological applications are required, which are easy to integrate with current practices, involve small capital investments, have quick payback periods, and add to farmer's commercial capability rather than foregoing current harvests for the benefit of future yields and the ecosystem.

Through reimagining the use of primary resources, principally soil, water and fossil fuels, and their interaction by reorganising traditional production models to use soilless technologies, such as aeroponics, it is possible to improve the food production output volume and nutritional quality while reducing water, fertiliser, pest control chemicals, and the use fossil fuels, or alternative energy sources, more efficiently. Soilless growing is not limited to traditionally cultivated land: through better use of available, and previously unusable suburban and urban land, as well as creating more space through growing vertically; effectively turning acreage into volume, soilless growing can bring poor quality land back into use and contribute towards the protection of soil based ecosystems. In addition to supporting the safeguarding of global forests, rainforests, and wetlands, this can have positive atmospheric environmental effects through reducing transportation and waste from 'field to fork'.

Waste within the agricultural system would be reduced through a production method change to aeroponic growing, transitioning from oversowing for contingency to 'growing to order', the ability to transport live produce to retailers or packaging wholesalers, a reduction in packaging if consumers take up a 'pick-your-own' model in mainstream retail outlets, and the ability to grow closer to the population centres who will consume the produce, meaning less perishes in transit post-packaging. Resultant waste can be post-processed into soil enhancements for soil restoration, and if utilising insects, can create inputs into fish or poultry food supply chains, in order to reduce the productive and environmental burden of producing meat.

Through urban farming, there are trickle-down effects in the creation of smart futures to a cascade of economies. In developed countries, towns are often centres of any unemployment and deprivation, and in developing nations, efforts to procure subsistence levels of nutrition and water can limit opportunities to run, operate, and innovate into other industries. Aeroponics offers opportunities to match unemployment with a business model that can payback over a short time frame and to create possibilities to access education and future employment away from subsistence.

If more produce agriculture can move towards urban areas, it becomes easier to justify protecting the ecosystems that are being eaten into through the current cultivatable land requirements. Although meat animal husbandry is a large factor in shifting agriculture, by utilising more biosecure and urban spaces, there may be commercial reason to support returning to a rotating grazing as part of a sustainable approach to rehabilitating soil.

Aeroponic growing can relieve pressure on land used for grain production, via a return to good crop rotation practices, with scope to invest in the soil's biome, through planting of cover crops to improve soil structure and nutrient profile and the reduced impact on the animals, insects, and microorganisms that abide there, through breaking the pesticide and till regime. Therefore, future harvests should not need as much energy invested through fertilisers, irrigations, and soil treatment chemicals to be productive, as are currently used through intensive farming practices. Commercially, with short payback periods and a wider range of crops able to be grown, aeroponic growing can be a truly food centred farm diversification model.

Ultimately, the challenge humanity faces is bigger than the transformation of agricultural practices. We need to transform the economic system that currently encourages all of us – producers, consumers, and investors – to act in an unsustainable way. Practices such as aeroponic growing facilitate the move towards change through its harmonisation with current consumer and retail demand practices.

References

Bentley, R. W. (2002) 'Global Oil & Gas Depletion: An Overview' Energy Policy Vol 30.

Buringh, P. (1989). Availability of agricultural land for crop and livestock production. In D. Pimentel & C. W. Hall (Eds.), *Food and Natural Resources* (pp. 69–83). San Diego: Academic Press.

Caribbean Regional Fund for Waste Water Management (2015) http://www.gefcrew.org/index.php/participating-countries/jamaica.

Cordell, D., Drangert, J.-O., & White, S. (2009). The story of phosphorus: Global food security and food for thought. *Global Environmental Change, 19*, 292–305. https://doi.org/10.1016/j.gloenvcha.2008.10.009. Retrieved 2013-09-10.

Glover, J. D., & Reganold, J. P. (2010). Perennial grains: Food security for the future. Issues in Science and Technology, 26(2).

Matthews, E., & Hammond, A. (1999). *Critical consumption trends and implications degrading earths ecosystems*. World Resources Institute. http://www.resilience.org/stories/2006-06-11/implications-fossil-fuel-dependence-food-system/.

Pimentel, D., & Giampletro, M. (1994). *Food, land, population and the U.S. economy*. Carrying Capacity Network. http://www.resilience.org/stories/2006-06-11/implications-fossil-fuel-dependence-food-system/.

Tanentzap, A. J., Lamb, A., Walker, S., & Farmer, A. (2015). Resolving conflicts between agriculture and the natural environment. *PLoS Biology, 13*(9), e1002242.

Vandermeer, J. (1995). The ecological basis of alternative agriculture. *Annual Review of Ecology and Systematics, 26*, 201–224.

World Water Assessment Programme (WWAP). (2009). *The United Nations World Water Development Report 3: Water in a Changing World*. Paris/London/Earthscan: UNESCO.

Chapter 11
Buildings that Perform: Thermal Performance and Comfort

Christopher Gorse, Martin Fletcher, Felix Thomas, Fiona Fylan, David Glew, David Farmer, and Pat Aloise-Young

Developers, designers and contractors are increasingly using titles such as 'green', 'eco' and 'low energy' to describe their buildings and reassure the environmentally conscious consumer of the green credentials of the property that they are investing in. Unfortunately, relatively few construction companies engage in research and development (R&D) to underpin their marketing rhetoric. The exceptions to this observation are those companies investing in innovation and engaging with both experts and customers to ensure a real understanding of property performance. Such activity is critical if the aspirational living and performance standards that are being claimed by the developers are to be attained and replicated on a wider scale. Forward-thinking developers are continuously using the lessons learned to inform and enable best practice, within their organisation, and through their exemplars setting standards across the construction industry.

Interest in the in situ performance of low-energy buildings has seen an increase in building performance testing, energy monitoring and occupant consultation with the aim of determining the various impact of low-energy technologies and construction approaches. Those companies advancing the industry are using research data to inform the way they build and develop low-energy properties and by appreciating the holistic context of the built environment are adding value to what may be construed as social sustainability.

Typically, companies will engage dedicated research establishments to undertake testing and monitoring to determine how buildings respond to their environment and ascertain how easily a building can be controlled to meet the occupant's needs. These tests and evaluations are thorough, exploring in forensic detail the achievements and issues within all aspects of a building, from the energetic performance of one element of building fabric to the combined influence of all systems on occupant satisfaction.

C. Gorse (✉) · M. Fletcher · F. Thomas · F. Fylan · D. Glew · D. Farmer · P. Aloise-Young
Leeds Sustainability Institute, Leeds Beckett University, Leeds, UK
e-mail: C.Gorse@leedsbeckett.ac.uk

© Springer International Publishing AG, part of Springer Nature 2018   193
M. Dastbaz et al. (eds.), *Smart Futures, Challenges of Urbanisation, and Social Sustainability*, https://doi.org/10.1007/978-3-319-74549-7_11

This chapter reports on the tests and monitoring undertaken on buildings, the common issues that they are designed to interrogate, and presents cases where developers are taking measures to achieve homes that function as expected and meet the requirements of the occupants, thus fulfilling their 'green' credentials.

11.1 Thermal Performance of the Built Environment

Before exploring the methods used to assess whether buildings are performing to their designed intention, it is relevant to consider the wider significance of building performance and why low-energy 'eco' buildings are gaining increased relevance in the construction industry.

The issues of anthropogenic climate change, fuel security and declining global fossil reserves require immediate efforts to be made to reduce our energy demand. In response to these global challenges, the UK government has made commitments to reduce greenhouse gas by at least 80% by the year 2050, relative to 1990 levels. The built environment plays a significant role in achieving this target. The housing sector is the UK's biggest annual energy user, accounting for 29% of the total energy consumption in the UK and approximately 30% of CO_2 emissions. The average household is estimated to consume 18,600 kWh of energy per year, of which 62% is used for space heating. The burning of natural gas for space heating accounts for around 35% of household CO_2 emissions (Palmer and Cooper 2013).

To reduce this consumption, building energy efficiency must be increased. The building stock in the UK is old (a situation that is also prevalent in the rest of Europe), with more than 40% of the dwellings being built before 1960 and 90% before 1990 (Artola et al. 2016). As the standards that require buildings to be energy efficient are relatively recent additions to building regulations, it is of no surprise that older buildings typically use more energy than modern buildings. Old buildings typically require more than twice the energy to heat when compared to modern buildings, yet with over 90% of buildings being more than 25 years old, inefficient older dwellings represent the vast majority of housing.

New buildings offer an easier route to energy reduction, being able to utilise modern materials and methods developed specifically for enhanced energy performance. However, new buildings either replace or expand the total stock at just less than 1% of the total buildings each year (with the building of new homes generally ranging from 140,000 to 250,000 per year). With current rates of construction together with an increasing population, meaning that 75–80% of the current building stock will still be in use in 2050 (Power 2008), it is essential that existing buildings are addressed; however, there is a considerable amount of refurbishment required to bring existing buildings up to modern energy-efficient standards (DCLG 2017; Artola et al. 2016).

The renovation rate of the UK's 28 million existing dwellings is low, with only 1–2% of the building stock renovated each year, with considerable variation in the level of upgrade (UK figures taken from: Valuation Office Agency 2016; National

Records of Scotland 2016; Eurostat 2017; Artola et al. 2016; DCLG 2017). Buildings are renovated differently, with some receiving superficial aesthetic upgrades, whilst others receive what has been termed 'deep retrofit', where the whole building fabric and services are upgraded to high-energy-efficient standards. Renovations that reduce both the delivered and the final energy consumption of a building by a significant percentage compared with the pre-renovation levels are insignificant in terms of national targets, representing less than 1% of all buildings renovated. Although the retrofit and upgrade market is small when compared to the whole building stock, construction activity related to renovation accounts for 57% of all building works.

Beyond the reduction of energy and greenhouse gas emissions, older buildings tend to be draughty, cold during the winter months and may experience some problems with condensation and mould, even in some cases affecting the health and wellbeing of the occupants. Evidence collected by the Leeds Sustainability Institute suggests many existing buildings are difficult to heat, with some occupants believing that such buildings had a negative impact on their family's health (Gorse et al. 2017a). The health risks to the occupants in some cases are acute, evidenced by the 40,000 excessive winter deaths in the UK, 9000 of which are associated with cold homes (ACE 2015; NEA 2016). During a 5-year period, up to 2015, there were 46,716 deaths attributed to cold dwellings. The annual death rate caused by cold and damp buildings is similar to that caused by alcohol and almost as high as breast cancer (ACE 2015). However, mortality is a particularly weak indicator of the impact of cold homes on the occupant; the years of healthy life lost and illnesses related to living in cold damp environments have much wider impact on the society; ill health caused as a result of living in damp, cold and draughty conditions has been estimated to cost the NHS £1.36 billion per year (NICE 2015).

Thermal upgrades have the potential to significantly improve the performance of a building, reducing energy demand and enhancing occupant satisfaction. To achieve high-energy-efficient standards and comfort gains, there is a need to understand the nature of the building stock and what is necessary to improve existing buildings and enhance new developments. Research activity facilitates this process by ensuring new products, and construction methods achieve their intended purpose and performance (Table 11.1).

11.2   Thermal Performance of the Building Fabric

Thermal performance is a major factor in the energy efficiency of buildings and is a function of the interaction of several building fabric elements and their influence on heat loss mechanisms. Maintaining comfortable conditions within a building can require a great deal of energy input, depending on the thermal performance of a given building. On average, 62% of household energy consumption is for space heating – this means that good thermal performance, either in new build design or in existing building retrofit, represents major potential for the reduction of energy demand for heating and thus considerable reduction in CO_2 emissions. In addition

Table 11.1 Overview of renovation market share in the UK (Artola et al. 2016)

Renovation undertaken	Market share and example of renovation	Approximate costs per m²
Minor renovations	85% of the market: 1 or 2 measures (e.g. a new boiler) resulting in a reduction in energy consumption of between 0% and 30%	Average costs of €60/m²
Moderate renovations	10%: 3–5 improvements (e.g. insulation of relevant parts of the dwelling plus a new boiler) resulting in energy reductions in the range of 30–60%	Average costs of €140/m²
Extensive renovations	5%: In this approach, the renovation is viewed as a package of measures working together leading to an energy reduction of 60–90%	Average costs of €330/m²
Almost zero-energy building renovations	Negligible: The replacement or upgrade of all elements which have a bearing on energy use, as well as the installation of renewable energy technologies in order to reduce energy consumption and carbon emission levels to close to zero	Average costs of €580/m²

Note: current renovation represents less than 1% of the total building stock but accounts for over 50% of total building activity

to energy and CO_2 savings, there are also benefits to housing occupants: more comfortable internal conditions and reduced energy bills.

The majority of heat loss in buildings is through the plane elements: walls, roof, floor, windows and doors. The rate at which heat is lost through these elements is expressed as their thermal transmittance, or U-value (watts per square metre per degree Kelvin). A lower U-value represents lower heat loss and higher insulating properties. Achieving low U-values for elements of the building fabric is the first step in reducing the heat loss of a building. The primary method of reducing heat loss through the plane elements is through the application of insulation.

The external envelope can both act as a thermally resistant material, a medium that allows solar heat energy in, and a place to store heat energy, which can be released at a later time to smooth out the heating and cooling cycles (Fig. 11.1).

Household expenditure on space heating represents a significant portion of annual costs, and whilst the actual cost per household varies, reducing energy required for heating presents a desirable outcome for all occupants. In a recent study conducted by Leeds Beckett University, it was shown that for a small two-bedroom Victorian terrace house, the cost of heating the home could be reduced from £554 (where there was no thermal upgrade) to £206 (where a full deep retrofit had been applied) with annual CO_2e emissions associated with space heating reducing from 2.31 tonnes (no thermal upgrade) to 0.86 tonnes. The full thermal retrofit resulted in reduction of 63% heat loss through the fabric (Gorse et al. 2017) (Fig. 11.2 and Table 11.2).

This example illustrates the aggregate potential of multiple thermal upgrades, addressing several individual elements of the building fabric, i.e. the walls, glazing and loft. As previously noted, such deep retrofits are rare; it is more common for retrofits to be single interventions targeting specific aspects of the building fabric, i.e. the application of a single form of insulation. Whilst the example describes a retrofit, the principles for high thermal performance for new and existing buildings are largely the same – the key difference being the route to thermal upgrade being

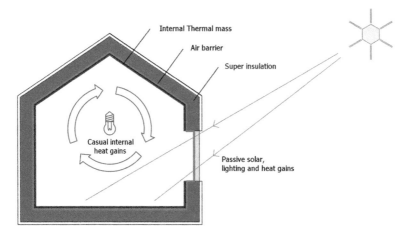

Fig. 11.1 Fabric first and passive approaches to creating energy-efficient and comfortable environments (Thomas and Gorse 2015; Gorse et al. 2016)

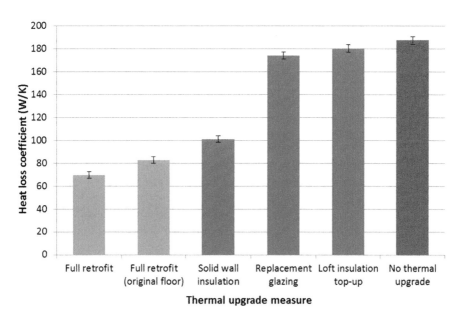

Fig. 11.2 Measured heat loss (heat loss coefficient HLC) of the test house at each test stage (blue bars represent the test house heat loss following a single thermal upgrade measure; green bars represent thermal upgrade measures in combination) (Gorse et al. 2017)

significantly more challenging for retrofit because of the limitations imposed by working with an existing building as opposed to the 'blank canvas' of a new build. In existing buildings, there are many parts of the external envelope that are difficult to access or treat. Internally, floors, cupboards and services will often disrupt internal insulation, whilst external insulation is affected by abutment, complicated detailing, services and junctions such as the eaves. Where the difficult to access areas remain uninsulated, they present a thermal weakness in the envelope.

Table 11.2 Impact of thermal upgrades on a Victorian two-bedroom end terrace in the North West of England (annual space heating demand and cost, and CO_2 equivalent emission reductions)

Thermal upgrade measure	HLC (W/K)	Reduction on baseline (W/K)	Annual space heating energy reduction (kWh)	Annual space heating cost reduction (£)	Annual space heating CO_2e reduction (kg)
Full retrofit	69.7	117.8	6497	348	1449
Full retrofit (original floor)	82.7	104.8	5777	310	1289
Solid wall insulation	101.2	86.4	4761	255	1062
Replacement glazing	174.2	13.4	737	39	164
Loft insulation	180.5	7.1	390	21	87
No thermal upgrade	187.5	n/a	n/a	n/a	n/a
Floor upgrade	n/a	13.1	720	39	161

Areas of lower thermal resistance than the surrounding building fabric are known as thermal bridges. Lower thermal resistance of the building fabric at thermal bridges leads to greater loss of heat through these parts of the building fabric than unbridged areas. This lower thermal resistance is often caused by a break in the insulation layer by a material of higher thermal conductivity than the insulation.

There are three kinds of thermal bridges that can be considered in a building: repeating thermal bridges, linear thermal bridges and point thermal bridges (Fig. 11.3). Repeating thermal bridges occur at regular intervals within the plane elements of the building fabric and are accounted for in the U-values of the element they are within. Linear thermal bridging often occurs along the length of junctions within the building fabric, due to the building geometry or detailing at these locations. Linear thermal bridging is represented by ψ (psi) values, in W/mK (watts per metre per degree Kelvin). Point thermal bridging occurs at isolated non-repeating locations within the building fabric, at a penetration through the building fabric or the location of a fixing. Point thermal bridging is represented by χ (chi) values, in W/K (watts per degree Kelvin).

When assessing the heat loss of a building, the total thermal bridging of all linear bridges and all point bridges are added together then divided by the total heat loss area of the building to give a y-value, in W/K (watts per degree Kelvin). As the thermal resistance of the plane elements of a building's fabric is increased through the addition of insulation, the heat loss through thermal bridges becomes a larger proportion of the total heat loss, unless consideration is given to mitigating thermal bridging (Fig. 11.4).

The basic principle of preventing thermal bridging is to ensure the continuity of the insulation layer throughout the fabric of the building envelope, though in practice breaks in the insulation layer are unavoidable, in almost all current building types. Where a bridge cannot be avoided, steps can be taken to reduce the extent of bridging. Reducing the amount of material bridging the insulation layer reduces the rate at which heat can be conducted over the thermal bridge. Additionally, introducing

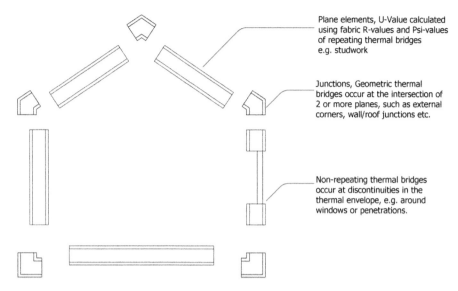

Fig. 11.3 Plane element, geometric thermal bridges and non-repeating thermal bridges (Thomas and Gorse 2015; Gorse et al. 2016)

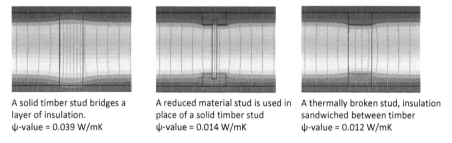

A solid timber stud bridges a layer of insulation.
ψ-value = 0.039 W/mK

A reduced material stud is used in place of a solid timber stud
ψ-value = 0.014 W/mK

A thermally broken stud, insulation sandwiched between timber
ψ-value = 0.012 W/mK

Fig. 11.4 Thermal bridges and ψ-values for stud sections (Thomas and Gorse 2015)

thermal breaks to the bridging elements, which involves using materials of high thermal resistance to isolate low thermal resistance elements, prevents a continuous bridge from the warm to cold side of the element.

In addition to heat lost via conduction through the fabric elements, the exchange of heated internal air with cold external air can be a significant source of heat loss from a building. In order to maintain a comfortable internal environment within a building, the air must be heated; however, air must also be refreshed through ventilation to prevent build-up of moisture, pollutants and odours that would make the internal environment unpleasant or unsafe. Heat loss through ventilation has two components: purpose-provided ventilation such as extractor fans, intended to maintain air quality, and uncontrolled ventilation through gaps in the building fabric, often referred to as air leakage. Exchange of warm internal air with cool external air represents a loss of heat from the internal environment of a building and a reduction in internal air temperature, requiring additional heat input.

Achieving low ventilation heat loss in buildings to reduce the volume of warm air lost and cold air entering the building is a key element of energy efficiency. Achieving a low air permeability requires that careful consideration be given to airtightness during design stages; a continuous airtight barrier should be specified and special consideration given to how the barrier will interact with junctions and openings.

Whilst a thermally efficient building fabric will reduce the energy required to condition the internal environment, it is not a complete solution and must be combined with appropriate low-energy systems for heating and ventilation. These in turn must be operated correctly by the occupant to ensure that predicted performance is achieved. A large component of the gap between the prediction of energy use and that actually used, 'the reality – in situ use', results from the interaction between systems and users; it is as important to explore the relationship between users and systems as is it to test the systems themselves.

To understand the building's performance and its characteristic behaviour, it is essential that the building fabric and services are tested and commissioned, and it is also informative to see how the building system responds under different occupation patterns. Once the buildings can be controlled, then it is possible to influence occupant behaviour to achieve optimum performance, based on the user needs; however, buildings must first perform. The first step to understanding buildings is to ensure that the designed performance is tested in the field, under real operating conditions. The following sections outline tests and monitoring.

11.3   Methods for Testing the Performance of Buildings

11.3.1   Fabric Testing

As previously stated, heat loss through the building fabric is significant in terms of energy demand and occupant satisfaction and is composed of two elements: heat loss via conduction through the plane elements and heat loss via uncontrolled ventilation. This has encouraged the development of many designs, products and systems to reduce the heat loss through the fabric and improve building airtightness. The development of these energy-saving innovations often occurs in a laboratory environment under unrealistic environmental conditions. The result of this is that whilst a building may meet targets for thermal performance in theory, a physically completed building may not reach its expected design performance when exposed to realistic conditions.

This shortfall in performance is referred to as the 'performance gap' and is caused by several factors such as incorrect construction, poor workmanship, failings at the design stage, poor detailing and improvisation on site (Johnston et al. 2015). It is common to find that new buildings are some way off their expected performance. Where buildings have been in use for some time and remedial action, refits or refurbishment has been undertaken, the builders' work and fitting of services are often incomplete, with penetrations through the fabric not being sealed. As well as affecting the building's thermal performance, such changes can impact on fire safety, allowing smoke and flames to breach compartmentation and spread through the fabric. If a fire takes hold of a building and such defects exist, the consequences for occupants can be deadly (Gorse and Sturges 2017). Thus, whilst the tests for thermal performance are useful, the forensic examination often reveals defects of the building that also have a potential impact on other aspects of performance.

It is important to undertake building performance evaluation to validate the fabric performance of finished buildings, demonstrate regulatory compliance and potentially locate faults leading to underperformance which can be corrected. In the case of multibuilding developments, findings of tests on early buildings can help remedy faults in later buildings before they are made.

It can be advantageous to undertake a number of tests concurrently, as the conditions required to undertake individual tests are also ideal for others. An intense programme of testing can thoroughly examine many aspects on a building's performance, helping to identify and remedy shortfalls in performance as well as inform future design and construction projects.

11.3.2 Coheating

The coheating test method is used to quantify the amount of heat energy a building loses through its fabric, in the form of a heat loss coefficient (HLC) expressed as W/K (watts per degree Kelvin). At its most basic, a coheating test is performed by heating a building to a set temperature, at least 10 degrees Kelvin above external air temperature, using electrical resistance heaters. The electricity required to maintain this continuous temperature difference is logged, and a HLC can then be calculated by plotting daily heat input (in kilowatt hours) against daily internal-external temperature difference (K). The gradient of the resulting plot gives the heat loss coefficient in W/K (Johnston et al. 2013). Taken by itself, the coheating test can only indicate the heat loss of the building fabric as a whole. Individual aspects of thermal performance must be investigated by other means, though a coheating test provides a good opportunity to carry out other tests concurrently (Fig. 11.5).

11.3.3 Heat Flux

Heat flux density measurement is used to measure the rate of heat flow into or out of a building element, effectively measuring the in situ U-value of an element that is tested. To perform heat flux measurement, a heat flux plate is placed on the element, avoiding any points of thermal bridging. During testing, heat flux density (W/m^2) is logged as well as air temperatures on either side of the element being tested so that when heat flux is divided by temperature difference, the resulting value is an effective U-value for the building element (W/m^2K).

Fig. 11.5 An example of a typical coheating setup: electrical resistance heater with thermostatic control to maintain constant internal temperature, electricity and temperature logging equipment and a circulation fan to ensure air temperature uniformity in the zone

Heat flux measurement is best undertaken with a constant elevated temperature on the internal side of the building element to ensure monodirectional heat flow from inside to outside for the duration of measurement. A coheating test provides ideal conditions to perform heat flux measurements; the effective U-values measured using heat flux tests will add to the findings of a coheating test.

11.3.4   IR Thermography

Infrared (IR) thermography uses specialised 'thermal cameras' that can detect the infrared band of the electromagnetic spectrum to visualise the temperature of objects and surfaces. IR thermography is often undertaken as a survey, examining a building to find temperature irregularities that may indicate a problem within building elements that could not be seen with the naked eye.

IR thermography should be carried out when the internal temperature of the subject building is elevated above external temperature; this will help make cold spots more apparent when viewed helping to identify building defects, such as air leakage paths or improperly fitted insulation. IR thermography is particularly useful for investigating thermal bridging, air leakage and moisture ingress, as these will cause temperature variations that are visible when viewed with a thermal camera. A coheating test provides an excellent opportunity to carry out IR thermography, as internal temperatures will be elevated to a homogenous temperature throughout a building. Caution should be exercised when reviewing IR thermography, as temperature differences can appear exaggerated or be understated if a camera is set to automatically set temperature range.

11.3.5 *Air Permeability and Ventilation Tests*

UK building regulations Part L1a require that new buildings undergo air permeability testing to demonstrate that they comply with the threshold air permeability of 10 m^3/(h•m^2) @ 50 Pa (an air leakage rate of 10 m^3 of air, per m^2 of building envelope, per hour at a pressure differential of 50 Pa). Allowances are made for large developments, permitting a sample of each dwelling type to be tested rather than every dwelling of that type. Knowing the air permeability of a building allows the rate of heat loss due to air leakage to be calculated.

Air permeability of a building is relatively quick and easy to test. The most commonly used method is the air pressure test, utilising a 'blower door' apparatus, using a controlled fan to depressurise the internal environment by blowing internal air out of the building and then pressurise the internal environment by blowing external air into the building. Purpose-provided ventilation is deactivated and sealed off before testing, as only uncontrolled, i.e. unwanted ventilation, should be measured. Airflow rate through the fan and pressure differential between the internal and external environments are recorded and plotted to calculate the air permeability at a pressure differential of 50 Pascale. Air permeability values from depressurisation and pressurisation can be averaged to account for the building's behaviour in both conditions.

A blower door test provides an opportunity to undertake air leakage path detection, as airflow and pressure differentials are increased above those expected during normal use. Detection of air leakage paths helps to guide remedial works where a building does not reach its target air permeability. During depressurisation, IR thermography can be used to locate paths of air movement into and around the building, as cold external air will be drawn into the building through gaps in the construction, cooling the building fabric. This cooling can be visualised with a thermal camera, thus identifying air infiltration paths. During pressurisation internal air is forced out gaps in the building construction; air movement can be detected using a handheld smoke generator, pinpointing the locations where internal air is escaping; the use of smoke allows air movement to be captured with a visual camera (Fig. 11.6).

11.3.6 *In-Use Monitoring*

Testing the building fabric performance under controlled test conditions using methods such as the coheating test is useful for providing benchmark figures of performance and understanding the physical capabilities of a building; however, the results do not necessarily reflect how a building will perform when occupied. In-use monitoring provides a way to measure energy consumption and performance of a building over time whilst occupied, giving a more realistic representation of building behaviour when exposed to transient heating and cooling cycles and variations of occupant behaviour.

Fig. 11.6 Thermogram captured during building depressurisation, showing the leakage path of cool air not shown by visual methods

In-use monitoring may be as simple as monitoring monthly energy metre readings to determine the influence of seasonal change on monthly consumption. More intensive in-use monitoring can allow a more complex analysis of energy use behaviour and the performance of a building. An intensive in-use monitoring programme may make use of electrical submetres on individual circuits or appliances within a building to record energy consumption for specific end uses, record temperature and air quality data in multiple zones within a building and record external weather data. Where renewable technology is fitted, the performance of these systems can also be monitored to assess their performance. Intensive in-use monitoring can be used to gather a large volume of data over a long time period, subsequently requiring a greater investment of time and effort to analyse, in addition to requiring a large amount of equipment. For this reason, intensive monitoring is best deployed in pilot studies or exemplar buildings.

In addition to capturing energy data, in-use monitoring also investigates the internal environment by monitoring temperatures, humidity and CO_2. The appreciation of the internal environment is essential in establishing the experience of the occupant in a space and allows objective judgements to be made regarding their comfort and predicted satisfaction which can be later validated by consulting with the occupant. The monitoring of internal conditions also allows this relationship between the internal and external environment to be established.

11.4   The Role of the Occupant

11.4.1   Occupant Behaviour

The performance gap is often attributed to errors in the building's design or construction or because insufficient detail leads to confusion or improvisation on site. There is an additional source: the occupant. Whilst there is much discussion around reducing errors in design and construction, occupant behaviour is often regarded as too complex to characterise and therefore neglected. However, a research to explore occupant behaviour has identified several ways in which occupant behaviour can easily be addressed or at least understood. Behaviour that reduces the building's performance is often caused by occupants simply not understanding how to use the building. Such behaviour can be misinterpreted as wilfully misusing the features of their home, which leads to under- or overheating. Instead, such behaviour can often be because occupants did not receive a sufficient (or indeed any) handover of their home and so did not receive enough explanation of how their behaviour in the home influences its energy efficiency (Linden et al. 2006; Isajsson 2014) or any discussion of how their lifestyle needs could be met by the energy efficiency features in their home. Often, contractors can assume that disruption to occupants should be minimised. In practice, this can mean that occupants do not understand how best to interact with their home to maximise energy efficiency and comfort. This can lead to both over- and underheating. This can be addressed by producing a checklist of conversations to have with occupants about using the heating, cooling and ventilation features in their home.

One such example of this was observed during monitoring of the GENTOO Passivhaus development in Sunderland (Fletcher et al. 2017). The monitored dwelling is of high thermal performance, and following energy monitoring was shown to achieve its energy and carbon targets. However, the dwelling experienced significant overheating throughout the year. This was due in part to the occupants not engaging the summer bypass function of the mechanical ventilation with heat recovery (MVHR) system, in addition to limiting their window opening behaviour as they were advised that this limits system efficiency. Following re-education on the systems within the property temperatures returned to a more comfortable level. This example serves to illustrate that even when all energy and carbon objectives have been satisfied, occupancy can have a significant impact on how we view the success of high-specification buildings (Fig. 11.7).

Another aspect to consider is that the occupants' needs change over time. If this occurs around the time of retrofit, any in-use monitoring can indicate a performance gap, whereas in reality, the home is performing as designed, but the occupants are using more energy. In a retrofit study conducted by Leeds Sustainability Institute involving external wall insulation (Gorse et al. 2017b), many of the occupants experienced lifestyle changes in which affected their energy behaviour. Common changes included a family having a new baby, so they heat their home to a higher temperature, both during the day and the night. Retirement often means that the home is heated during the day as well as in the evening. Several of the homes in the study had additional people moving into or out of the home, so that the rooms are used to a greater or lesser extent. Illness could also mean that the home is heated or used in a different way. All these factors affect energy use, and whilst designers and construction companies would not want to stop occupants from changing how they use their home, any building performance assessment should include work to identify and understand such changes. This can be incorporated into a post occupancy evaluation (POE) questionnaire or qualitative work in the form of interviews or focus groups.

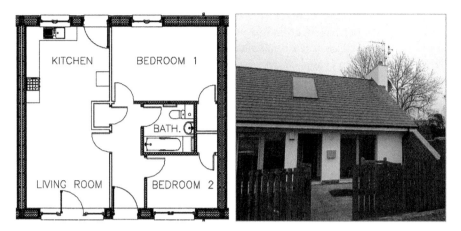

Fig. 11.7 GENTOO Passivhaus floor plan and front elevation

As well as understanding how occupants influence the performance gap, research with occupants can provide evidence on how housing influences quality of life. Living in a warmer more comfortable home could potentially lead to people feeling happier, and it has the potential to improve any long-term conditions they have.

Buildings should both be designed for energy efficiency and be capable of responding to the expectations and demands of the occupant. Detailed research observing occupant behaviour and building performance can help to understand and achieve optimum performance.

11.4.2 Post Occupancy Evaluation (POE)

Ultimately, occupants will behave in such a way as to ensure their own personal satisfaction even if this means bypassing the efficient technologies available to them and incurring an energy or carbon cost. As such, a sustainable development should aim to have minimal negative influence on the daily lives of the occupant. Therefore, in addition to energy monitoring for validation of performance, it is important to conduct post occupancy evaluation of novel designs and systems to gauge the impact on the user. Adaptive behaviours are significant when considering personal factors of thermal comfort, wellbeing and tolerance – as noted, if a sustainable technology places negative pressure on the occupant, then the likelihood is that either the system will be bypassed (thus negating its positive environmental impact) or the occupant will have an undesirable experience. The gap in performance between prediction and reality is well researched in the building fabric context (Johnston et al. 2015); however, it is equally important to evaluate the veracity of assumptions made with regard to the user and their behaviour within an environment as such assumptions underpin in-use performance predictions. POE is one such method to capture this information.

The most commonly used form of POE is the distribution of feedback questionnaires. The ubiquity of computers and smartphones has made the distribution, collection and analysis of online questionnaires increasingly cheap and simple. The Building Use Studies (BUS) methodology is an established example of this form of POE, comprising a three-page structured questionnaire which presents the occupant with various questions relating to the design, lifestyle and comfort of their building (Leaman 1995). Respondents are presented with various scales of different form and style which are specifically designed to deter feedback fatigue, in addition to several opportunities to enter their own comments. The data gathered is then compared with benchmarks generated from a database of exemplar buildings, with scores awarded and graphically represented in a 'traffic light' format for each category.

Whilst able to gather valuable information, the rigid format of structured questionnaires limits the possible richness of data by inadvertently establishing boundaries around how opinions may be expressed within the context of a specific question or topic group. For a comprehensive understanding of occupant experience, interviews and focus groups offer a more flexible approach and however come with increased complexity in terms of participation, cost and analysis.

The recommendations for what to include when designing research on occupant behaviour, based both on previous research and experience of researching the impact of occupants on energy use, are shown below:

- Who lives in the home (people and pets) and any regular visitors, such as grandparents providing childcare
- How the home is used, e.g. when people are in and out of the home, whether rooms are used differently in winter and summer, whether windows are kept open and doors are opened frequently
- Any health problems that people have that might affect the temperature of the home and the energy they use, together with a measure of their health status
- Preferred temperature within the home, the reason for this and reasons why the actual temperature is not the same as the preferred temperature
- Any life events that mean people might change the energy they use, such as spending more or less time in the home, having a baby, losing their job or retiring
- Understanding of how to use the heating and ventilation systems
- Confidence in the ability to use less energy ('perceived behavioural control')
- Beliefs about the advantages and disadvantages of using more or less energy
- Beliefs about what other people expect in terms of energy use ('social norms')
- How satisfied people are with factors such as how quickly their home heats up, how warm it gets, how draughty it is, how damp it gets, how much it costs to heat and how much noise it lets in
- How satisfied people are with living in their neighbourhood, for example, its appearance, how safe it feels and how much they feel they belong

11.5   Buildings that Perform

The acknowledgement of the role the built environment plays in global energy consumption has encouraged significant efforts to reduce the ecological footprint of buildings from both an energy and carbon perspective (UNEP, 2012). There are over 70 definitions for low- or zero-energy/carbon buildings in use globally (Williams et al. 2016), adding complexity to adoption on a wider scale and resulting in the creation of several assessment methods designed to facilitate environmentally conscious construction whilst offering differing views on what constitutes success. The foremost example in the UK is Part L of the building regulations (DCLG 2010, 2013) which emphasises the reduction of energy through improvement in the building fabric. Part L requires certain targets for building airtightness and fabric performance to be fulfilled, with compliance verified by the Standard Assessment Procedure (SAP) (BRE 2016). Whilst significant due to its legal status, Part L does not result in what may be termed as low-energy buildings, operating more to ensure an enforced lower limit on building energy performance. Described below are examples of buildings designed to energy performance standards that go beyond traditional requirements to achieve what may be fairly described as 'green', 'eco' or 'low energy'.

11.5.1   Passivhaus and EnerPHit

The 'fabric first' logic present in Part L has been extended with the creation of the Passivhaus standard (Feist et al. 2005). Developed by Wolfgang Feist and Bo Adamson, the Passivhaus approach is to create a building with exceptional thermal performance and airtightness, with ventilation controlled by a mechanical ventilation with heat recovery (MVHR) system. This, coupled with a design to maximise both solar and additional incidental heat gains, allows the heating energy demand to be minimised. One example of a successful Passivhaus project in the UK is the Denby Dale Passivhaus in West Yorkshire (Fig. 11.8), which was the first building in the UK to receive Passivhaus certification using traditional cavity wall construction. The dwelling achieved the goal of using 90% less energy for space heating compared to the UK average (Green Building Store 2017). Due to the strict requirements for certification, Passivhaus construction is more suited to new build developments. The principles, however, may still be applied to existing buildings in a retrofit. The EnerPHit standard has been developed for this purpose, acknowledging the challenges of existing buildings and providing slightly relaxed targets whilst still following Passivhaus principles to ensure a significant thermal performance improvement.

Fig. 11.8 Denby Dale Passivhaus (Green Building Store 2017)

11.5.2 Nearly Zero-Energy Buildings (nZEB)

Whilst reducing the space heating energy demand of buildings, the fabric first approach in isolation does not offer a solution to the energy required for systems and appliances within the building. Nearly zero-energy buildings (nZEB) as defined by the Energy Performance Buildings Directive (European Union 2010) seek to address these in-use energy considerations, combining high-performance building fabric with on-site renewable energy generation to provide full building energy requirements. The aspiration is an annual net zero-energy cost, i.e. over the course of the year; the building will generate at least as much energy as it consumes.

The starting point of nZEB is to maximise operational efficiency to reduce in-use energy demand. Fabric first construction principles are combined with energy-efficient services to minimise energy use within the building. Heat pump systems are typically used in combination with MVHR to minimise energy for heating. Lighting and appliances within the building will also be low energy, and all systems will typically be powered electrically to maximise the energy generated by on-site renewables.

Once efficiency measures have been implemented, the result is a smaller demand which can feasibly be met by on-site generation. Due to peak energy demand and supply seldom occurring simultaneously and the complications of efficient energy

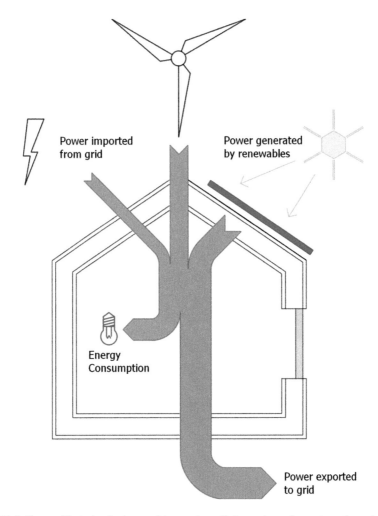

Power imported
from grid

Power generated
by renewables

Energy
Consumption

Power exported
to grid

Fig. 11.9 Renewable technologies used to meet small demand requirements and create nZEB buildings (Thomas and Gorse 2015; Gorse et al. 2016)

storage, nZEB buildings typically maintain a connection to the electric grid network. This allows electricity to be exported to the grid during periods of excess supply and imported when supply is low, ultimately balancing over the course of the year. Whilst theoretically sound, the success of nZEB is extremely sensitive to occupant effects, with correct operation of efficient systems essential if projected energy use is to reflect operational reality (Fig. 11.9).

11.5.3   BREEAM

When considering sustainable construction, the focus is predominantly on the reduction of energy and carbon emissions, with success measured by their reduction against predefined levels. It is equally important, however, to consider the wider impacts of construction beyond these two elements. The Building Research Establishment EnergyAssessment Method (BREEAM) is one such assessment designed to capture the holistic factors of the built environment, covering not just energy and carbon but also water, waste and transport amongst other things (Fig. 11.8).

Energy	Health and wellbeing	Innovation	Land use	Materials
Management	Pollution	Transport	Waste	Water

BREEAM assessment categories (BRE 2017a)

Launched in 1990, the BREEAM method has grown to become one of the most widely accepted assessment methods globally, with over 562,000 certified developments across 78 countries (BRE 2017a). The BREEAM method considers the above categories with weighting for their significance, awarding points for good practice which are aggregated and used to generate an overall percentage score, translating to an award from unclassified (<30%) to outstanding (85% or greater). BREEAM may be applied to both new build and retrofit constructions, and one such example is the 119 Ebury Street development in London, which achieved an 'outstanding' score. This project involved the conversion of a grade two listed building previously functioning as a hotel into three duplex apartments (BRE 2017b).

Internal wall insulation and triple-glazed windows improve the existing building fabric whilst maintaining the external aspect of the building. This is supplemented by the addition of phase change materials in the upper and middle floor apartments to help stabilise internal temperatures (UK-GBC 2017), absorbing excess heat and releasing it slowly during cooler periods. Heating provision comes from modern boilers with flue gas heat recovery and is supplemented by solar thermal panels. The efficiency of the heating system is further improved with the inclusion of an MVHR system to minimise space heating requirement Photovoltaics provide electrical energy to the first floor apartment, and there are several other features designed to minimise the environmental impact of the building, such as a rainwater harvesting system.

In addition to the provision of efficient fabric and services, 119 Ebury Street also acknowledges the role that occupants play in determining the success of a sustainable building. The apartments are equipped with a 'smart home' system to provide feedback on energy use and limit poor energy behaviour such as heating spaces with open windows. The understanding and correct operation of low-energy systems is essential to their success; however, they are often complex and unfamiliar, leading to misuse which ultimately negates their positive impact.

11.6   Conclusion

There are many aspects of performance which can be monitored and measured to ensure the building performance desired is achieved. For those wanting to develop high-performing buildings, a systematic approach to research and development, iteratively improving design and build, is required. However, without obtaining feedback and data on the building's performance, it is difficult to identify which aspects of the building perform and where buildings require improvement. This chapter has provided a context to the building stock and why it should be improved and provided tools for monitoring and measuring to ensure the performance of buildings can be improved.

References

ACE. (2015). *Chilled to death: The human cost of cold homes. London: Association for the conservation of energy.* http://www.ukace.org/wp-content/uploads/2015/03/ACE-and-EBR-fact-file-2015-03-Chilled-to-death.pdf. Accessed online 1 July 2017.

Artola, I., Rademaekers, K., Williams, R., & Yearwood, J. (2016). *Boosting building renovation: What potential and value for Europe?* http://www.europarl.europa.eu/RegData/etudes/STUD/2016/587326/IPOL_STU(2016)587326_EN.pdf. Accessed online 1 July 2017.

BRE. (2016). SAP: The Government's standard assessment procedure for energy rating of dwellings 2016, Building Research Establishment, UK.

BRE. (2017a). *BREEAM assessment categories.* http://www.breeam.com/. Accessed online 22 Aug 2017.

BRE. (2017b). *119 Ebury Street development.* http://www.breeam.com/index.jsp?id=800. Accessed online 22 Aug 2017.

DCLG. (2010). *Approved document L1B: Conservation of fuel and power in existing dwellings: 2010.* Edition, Department for Communities and Local Government, London.

DCLG. (2013). *Approved document L1A: Conservation of fuel and power in new dwellings: 2013.* Edition, Department for Communities and Local Government, London.

DCLG. (2017). *Statistical data set: Live table on house building: New build dwellings.* Department for Communities and Local Government, London. https://www.gov.uk/government/statistical-data-sets/live-tables-on-house-building. Accessed online 22 Aug 2017.

European Union. (2010). *Directive 2010/31/EU of the European Parliament and of the Council of 19 May 2010 on the energy performance of buildings.* http://eur-lex.europa.eu/legal-content/EN/ALL/;ELX_SESSIONID=FZMjThLLzfxmmMCQGp2Y1s2d3TjwtD8QS3pqdkhXZbwqGwlgY9KN!2064651424?uri=CELEX:32010L0031, Accessed online 22 Aug 2017.

Eurostat. (2017). *Housing Statistics, Eurostat Statistics Explained.* http://ec.europa.eu/eurostat/statistics-explained/index.php/Housing_statistics. Accessed online 22 Aug 2017.

Feist, W., Schnieders, J., Dorer, V., & Haas, A. (2005). Re-inventing air heating: Convenient and comfortable within the frame of the passive house concept. *Energy & Buildings, 37*, 1186–1203.

Fletcher, M. J., Johnston, D. K., Glew, D. W., & Parker, J. M. (2017). An empirical evaluation of temporal overheating in an assisted living Passivhaus dwelling in the UK. *Building and Environment, 121*, 106–118.

Gorse, C., Thomas, F., Glew, D., & Miles Shenton, D. (2016). Achieving Sustainability in New Build and Retrofit: Building Performance and Life Cycle Analysis. Chapter 8. In M. Dastbaz, I. Strange, & S. Selkowitz (Eds.), *(2015). Building sustainable futures: Design and the built environment* (pp. 183–207). Switzerland: Springer.

Gorse, C., Farmer, D., & Miles Shenton, D. (2017a). *Upgrading the Building Stock: Insights into the Performance of Ready for Market Solutions.* Presented at the International Refurbishment Symposium: Productivity, Resilience, Sustainability. Eco Connect London, Friday 15th September 2017 Institution of Engineering and Technology, London.

Gorse, C., Glew. D., Johnston, D., Fylan, F., Miles-Shenton, D., Smith, M., Brook-Peat, M., Farmer. D., Stafford. A., Parker. J., Fletcher, M., & Thomas, F. (2017b). *Core cities Green Deal Monitoring project.* Prepared for Department of Energy and Climate Change. DECC. Leeds Sustainability Institute, Leeds Beckett University.

Gorse, C., & Sturges, J. (2017). Not what anyone wanted: Observations on regulations, standards, quality and experience in the wake of Grenfell. *Journal of Construction Research and Innovation, 8,* 72–75.

Green Building Store. (2017). *Denby Dale Passivhaus.* https://www.greenbuildingstore.co.uk/technical-resource/denby-dale-passivhaus-uk-first-cavity-wall-passive-house/. Accessed online 27July 2017.

Isajsson, C. (2014). Learning for lower energy consumption. *International Journal of Consumer Studies, 38,* 12–17.

Johnston, D., Miles-Shenton, D., Farmer, F., & Wingfield, J. (2013). *Whole House Heat Loss Test Method (Coheating).* Leeds Metropolitan University. http://www.leedsbeckett.ac.uk/as/cebe/projects/iea_annex58/whole_house_heat_loss_test_method(coheating).pdfhttp://www.leedsbeckett.ac.uk/as/cebe/projects/iea_annex58/whole_house_heat_loss_test_method(coheating).pdf. Accessed online 27 July 2017.

Johnston, D., Miles-Shenton, D., & Farmer, D. (2015). Quantifying the domestic building fabric 'performance gap'. *Building Services Engineering Research and Technology, 36*(5), 614–627.

Leaman, A. (1995). *Building Use Studies Methodology.* http://www.busmethodology.org.uk/history/. Accessed online 22 Aug 2017.

Linden, A.-L., Carlsson-Kanyama, A., & Eriksson, B. (2006). Efficient and inefficient aspects of residential energy behaviour: What are the policy instruments for change? *Energy Policy, 34,* 1918–1927.

National Records of Scotland. (2016). *Estimates of household and dwellings in Scotland 2016.* https://www.nrscotland.gov.uk/files//statistics/household-estimates/2016/house-est-16.pdf. Accessed online 22 Aug 2017.

NEA. (2016). *Get warm soon? Progress to reduce ill health associated with cold homes in England.* London: National Energy Action.

NICE. (2015). *Costing statement: Excess winter deaths and illness. Implementing the NICE guidance on excess winter deaths and illnesses associated with cold homes (NG6).* National Institute for Health and Care Excellence, March 2015. https://www.nice.org.uk/guidance/ng6/resources/costing-statement-pdf-6811741. Accessed online 1 July 2017.

Palmer, J., & Cooper, I. (2013). *United Kingdom Housing Energy Fact File.* Department of Energy and Climate Change, TSO, London.

Power, A. (2008). Does demolition or refurbishment of old and inefficient homes help to increase our environmental, social and economic viability? *Energy Policy, 36,* 44874501.

Thomas, F., & Gorse, C. (2015). *An illustrated review of the factors contributing to differences between designed performance and in-use energy demand.* LSI Technical Paper 2015-TP-0001. Leeds Sustainability Institute. http://www.leedsbeckett.ac.uk/as/cebe/resources.htm. Accessed online 3 Sept 2017.

UK-GBC. (2017). *Case Study on-site learning: 119 Ebury Street.* http://www.ukgbc.org/resources/case-study/case-study-site-learning-119-ebury-street. Accessed online 22 Aug 2017.

UNEP. (2012). *Building design and construction: Forging resource efficiency and sustainable development.* In: M. Comstock, C. Garrigan, S. Pouffary (Eds.), United Nations Environment Programme.

Valuation Office Agency. (2016). *UK Statistical Release notes. Council Tax: Stock of Properties England and Wales, 2016.* https://www.gov.uk/government/uploads/system/uploads/attachment_data/file/540756/CTSOP_160630_statistical_release_notes.pdf. Accessed online 3 Sep 2017.

Williams, J., Mitchell, R., Raicic, V., Vellei, M., Mustard, G., Wismayer, A., Yin, X., Davey, S., Shakil, M., Yang, Y., Parkin, A., & Coley, D. (2016). Less is more: A review of low energy standards and the urgent need for an international universal zero energy standard. *Journal of Building Engineering, 6*, 65–74. ISSN 2352-71.

Chapter 12
Maintaining Excellence and Expertise Within Medical Imaging: A Sustainable Practice?

Christopher M. Hayre

12.1 Introduction

This chapter explores sustainable futures within the medical imaging profession. It begins by introducing healthcare professionals, diagnostic radiographers and diagnostic radiologists. This is important because they are used interchangeably throughout this chapter when discussing issues concerning the profession. Next, the author will examine what the term 'sustainability' means for the medical imaging profession. It is important to critically assess the term, how it is used currently and what original discussions can be applied in order to offer a critical examination of current challenges and opportunities faced within the medical imaging profession. Three overarching themes are explored and justified in ensuring sustainable futures of the medical imaging profession.

First, the author explores advancing technology in contemporary radiographic practices because whilst there are widely accepted advantages following the introduction of computing hardware, the author discusses limitations that may resonate with practitioners nationally and internationally. Further, the suboptimum use of ionising radiation is uncovered, whereby technological enhancements have not only led to increases in radiation doses to patients but may also be unnoticed by practitioners delivering ionising radiation to patients.

Second, the author outlines existing challenges and opportunities within the radiography curriculum. He begins by arguing that radiographers should have sufficient knowledge and understanding of advancing technology in order to optimise it appropriately. Further, the cessation of health degree funding to healthcare courses nationally (including the author's institution) warrants an opportunistic discussion regarding a higher education institutions (HEIs) business model, critically

C. M. Hayre (✉)
University of Suffolk, Ipswich, United Kingdom
e-mail: c.hayre@uos.ac.uk

© Springer International Publishing AG, part of Springer Nature 2018
M. Dastbaz et al. (eds.), *Smart Futures, Challenges of Urbanisation, and Social Sustainability*, https://doi.org/10.1007/978-3-319-74549-7_12

challenging whether existing health degree courses remain the appropriate products for prospective students embarking onto a health career.

Third, the author discusses a recently debated issue concerning radiographer reporting. Currently, professional bodies aligned to both diagnostic radiography and diagnostic radiology offer juxtaposed statements regarding the progression of image interpretation by radiographers, a role previously undertaken by radiologists. In addition, the author recognises the importance of exploring other areas of 'sustainable futures' by offering researchers a research methodology that fits to the role of health practitioners clinically.

12.2   Medical Imaging: The Profession and Practitioners

Ever since the discovery of X-rays by Röntgen in 1895, the practice of medical imaging to uncover suspected pathology has continued to develop worldwide. The practice of medical imaging is primarily encompassed within the hospital environment, upon which this chapter focuses. The term 'medical imaging' is used throughout this chapter but is known by other taxonomies, including diagnostic radiography and diagnostic radiology. The former is linked to healthcare professionals (diagnostic radiographers) that are educated to degree level to acquire radiographic images using a number of imaging modalities for image interpretation. The latter are associated with diagnostic radiologists who are medically educated professionals that have chosen to 'specialise' in interpreting medical images from an array of imaging modalities, which can take 12–15 years of study. These roles are important to identify, as they will be discussed interchangeably.

In the United Kingdom (UK), medical imaging examinations continue to increase and encompass a range of imaging modalities, including general imaging, computed tomography (CT), magnetic resonance imaging (MRI), ultrasonography, fluoroscopy, mammography and angiography (Health Protection Agency 2011). This chapter explores sustainable futures within general radiography because it approximately constitutes 90% of all medical imaging examinations in the UK (ibid). By using empirical research and evaluating contemporary challenges and opportunities, it will enable a reader to gain valuable insight into a large segment of the medical imaging profession and offer discussions that remain paramount to ensure the sustainability to the profession and of patient care.

The author is an educator, practitioner and researcher within the diagnostic radiography profession. His positionality is supported by an auto/biographical reflexive lens, supported by empirical research undertaken from 2012 to 2013, which examined diagnostic radiography practices in two imaging departments within the UK.

12.3   Sustainability: What Does This Mean for Medical Imaging?

This chapter explores three themes pertinent to the sustainable futures of diagnostic radiography: advancing technology, radiographic education and professional cultures. Before these are examined, it is important to question what 'sustainable futures' represents for medical imaging. This is important for two reasons. First, by offering a definition from the author's position, readers can identify his own professional viewpoint (as a diagnostic radiographer). Second, the author accepts that his position on the topics discussed may be contested, as it remains a perspective of a possible many.

So, what is sustainability and what does it mean for the medical imaging profession? At first, it was believed that this would be a relatively easy question to answer; yet much disparity of the term 'sustainability' exists in the literature. Currently, literature within diagnostic radiography recognises sustainability as a way of minimising ionising radiation to patients and staff members due to the potential occurrence of stochastic and deterministic biological effects, which can lead to cancer induction (Picano 2004). This is discussed later and remains pertinent to the general imaging modality. Furthermore, a recent report titled 'creating a sustainable workforce' (SoR 2016) examines the potential dangers following the cessation of government funding for undergraduate radiography programmes in HEIs. This will be explored in Sect. 5.5.2. In diagnostic radiology, the Royal College of Radiologists (RCR)) offers a number of sustainable publications, focusing upon the 'lifelong learning of radiologists' (RCR 2016), 'working for alternate providers', (RCR 2015a), 'flexible home working' (RCR 2015b) and a way of avoiding 'a radiology crisis' (RCR 2017) by rejecting role advancements of radiographers (discussed in Sect. 5.6.1). Whilst considering current topical issues, the author also offers previously underexplored tenets of sustainability within diagnostic radiography, which remain central when considering sustainable futures of healthcare practices. Few texts offer wider social-cultural discussions regarding sustainable futures that can facilitate and/or hinder the development of radiographic healthcare. Further, such insights may not only provoke wider discussion amongst the medical imaging community; however, they may be transferable to other health disciplines.

Authors outside of medical imaging profession define sustainability as 'meeting the needs of the present without compromising the ability of future generations to meet their own needs' (Jacques 2015, p. 19). Whilst widely cited, Buchanan et al. (2007) offer an alternate definition following their study exploring senior manager's perspective of sustainability within the National Health Service (NHS). Buchanan et al. (2007, p. 30) definition goes some way of accepting its multifaceted importance within the health and social care community, defining it as 'maintaining a process of quality improvement, commitment and leadership, improving techniques, education and training, teamwork and cultural change'. In short, this affirms that not only is sustainability pertinent to diagnostic radiography and other professions in the NHS, however, it can involve a plethora of topics, ranging from practitioner competencies, advancing technology, patient care and interprofessional working. This supports the rationale for this chapter to examine and accept advancing technology, radiography education and professional cultures as central components to sustainability in medical imaging.

The topics discussed within these themes are threefold. First, following the recent advances in radiographic technology, the author critically evaluates whether advances in technology remain the panacea for the radiography profession by challenging whether existing approaches are sustainable for the profession. Areas of discussion include the 'technology push/pull' theory and how it may benefit and/or hinder patient experiences. Further, the use of ionising radiation by practitioners is discussed. This is important to consider within general radiography as improper use of ionising radiation can lead to increased risks of cancer induction to patients. Second, radiographic education will be discussed with a focus on challenging whether radiographers remain image acquisition or image interpretation experts following recent advances of role progression of radiographers. In addition, following the cessation of financial stability amongst HEIs for health programmes nationally, the author offers a business model that challenges course delivery whilst critically asking whether contemporary undergraduate courses meet the student's needs and challenging whether higher education requires a pedagogical paradigm shift. Lastly, the author discusses the importance of embedding sustainability into our culture, values and beliefs by considering anthropocentrism as an overarching model, supported by deontological and utilitarianism principles, whereby patients remain at the centre of the medical imaging profession. Further, the importance of undertaking further research enhancing knowledge and understanding is recognised, with the use of an innovative methodology utilising ethnography. This is proposed because it may provide healthcare practitioners with a unique insight into sustainable futures. It is claimed that by using qualitative and quantitative methods within a single 'umbrella methodology', healthcare practitioners can apply similar methodologies enabling a critical examination of their own practices and sustainability of healthcare delivery.

In response to Sect. 5.3, it is apparent that the sustainability of health and social care remains multifaceted and therefore cannot simply be a 'one-size-fits-all' methodology. Medical imaging has many subspecialties, encompassing the complex nature of the various imaging modalities and practices. What is generally accepted is that the term 'sustainability' remains pertinent to the medical imaging profession and can enable practitioners to critically assess current healthcare practices and cultural norms and ensure that optimum healthcare is maintained now and for the future.

12.4   Advancing Technology: The Panacea for Patient Care?

It is easy to see how advances in technology have altered individual lives, whereby communication, travel and business utilise technology in some form. Garetti and Taisch (2012) claim that technology is a major component of evolutionary change and acknowledges that technology can offer environmentally sustainable approaches for current and future generations. Medicine is not dissimilar and is becoming increasingly reliant upon technical equipment to diagnose and manage medical conditions (Becker et al. 1961; Fett 2000). One example within medical imaging is the removal of X-ray film following advancements of acquiring X-ray images [radiographs] digitally using amorphous selenium (a-Se) detectors. In the past, X-ray film processing was generally associated with environmental pollution, using photo chemicals, which were used to develop X-ray films prior to diagnosis. However, the introduction of digital radiography equipment negates the use of film, thus removing the use of pollutant chemicals. Digital detector classifications are related to the conversion of X-ray photons into an electric charge and can use a direct or indirect process to convert X-rays into electrical charges using a thin-film transistor (TFT) array. Direct conversion is becoming commonly associated within general radiography whereby detectors use an X-ray photoconductor converting X-ray photons directly into electrical charges. In addition to the formation of 'a digital radiograph', radiographs are now stored using a picture archiving communication system (PACS). A PACS is commonly used for electronic storage, retrieval, distribution, communication, display and manipulation of radiographs (and other imaging data) worldwide. In short, technological advances seen in the last two decades have revolutionised the medical imaging profession and have arguably facilitated sustainable futures following improvements of ergonomics, access, space and reduction of pollutant chemicals.

Whilst technology has alleviated many historical problems associated with medical imaging practices, patient care and use of ionising radiation will be explored as there may be limiting factors previously underexplored within digital radiography. This is important because whilst reports continue to demonstrate the benefits of advancing technology, it is important to reflect clinically on the use of technology by radiographers and how it may impact healthcare professions at a local level.

12.4.1 Technological Push/Pull Theory: Does Technology Really Enhance Patient Care?

A patient is a person that receives healthcare services and is often in an ill or injured state and in need of treatment by a physician, nurse and/or allied health professional. The word 'patient' derives from the Latin word 'patiens', meaning 'I am suffering' and akin to the Greek verb 'to suffer'. It is generally accepted that patients are at the centre of healthcare delivery in the NHS whereby care is delivered to the highest standards of excellence and professionalism. This is advocated in the NHS Constitution in the UK (Department of Health 2015); yet following the introduction of digital radiography, few texts critically explore general radiographic examinations and how these may affect patients during their examinations.

Anthropologists, sociologists, historians and economists have repeatedly shown that technologies transform societies and altered relations within organisations

(Blau et al. 1976 cited in Barley 1986). Following the widespread recognition of the role of technology in economic growth, a debate during the 1960s and 1970s challenged whether the rate and direction of technological change was influenced by changes in market demand or by advances in technology. The theory of 'technology push' and 'demand pull' remains to be a catalyst for debate and is important to understand if we are to propose that this could impede sustainable futures within medical imaging (Nemet 2009). 'Technological push' is suggested to be a hindrance to practitioners because they often remain in passive roles, as information receivers whereby technology is often 'pushed' into clinical environments. The notion of 'market pull' is whereby a product remains hard to resist for radiology managers wishing to purchase a new piece of X-ray equipment. In diagnostic radiography, this resonates with the claim that digital radiography enhances the ease of radiological examinations and offers reductions in ionising radiation: 'exams are done easier and may result in fewer retakes and a low X-ray dose for your patients' (Philips 2011, p. 3). The challenge, then, for the medical imaging profession is whether these critiques impact on patient care during X-ray examinations.

During the author's research in two acute hospital settings, a central focus aimed at exploring whether advancing technologies affected patient experiences. Two aspects were uncovered by the author's research. On the one hand, advantages of digital radiography were due to the immediate visibility of the digital radiograph. The introduction of digital radiography enabled radiographers to be present at all times with their patients. In contrast to using film, whereby radiographers were often required to leave patients alone in the X-ray room in order to 'process/develop' films for clinicians (Hayre 2016a, pp. 160–161):

> You don't have to leave the room, you don't have to disappear. You're there the whole time with your patient. And I think from that point of view that is better for the patient. (Senior radiographer)

> Because of you not abandoning the patient in the room. Especially nervous patients, [they] aren't being left alone in a scary room. So it does help that rapport with the patient. You see the images straight away, so you can help reassure patients quickly. (Senior radiographer)

Here, senior radiographers offer clear advantages over conventional X-ray film, which required radiographers to process and develop radiographs in a 'processing environment'. This offers insight by highlighting that advances in technology may have a direct impact on patient care, whereby patients are observed at all times during their care, and arguably facilitating patient experiences. Yet, on the other hand, radiographers from the same study commented on what they felt was inferior care upon using digital radiography equipment. Radiographers commented on the lack of versatility with the equipment, which in their view impacted on patient experiences when undertaken radiographic examinations (Hayre 2016a, p. 163):

> I think sometimes, when we try and shoehorn a patient into digital radiography it makes the experience worse for the patient. Sometimes we try and use digital radiography – we use the horizontal bucky and somebody's broken their arm or their elbow, and you sort of yank them over and you think "You will get the image on digital radiography! When you could almost not move the patient at all. There are instances when we're yes, shoehorning a

> patient into digital radiography is actually reducing the patient experience [sic]. And it makes it more unpleasant for the patient sometimes. (Radiographer)

> It's horrible to say, but you try not to cause them any pain but if you've got to get the best image possible, you're going to have to move them… And you don't want to keep getting them back up when doctors keep saying, "that's not good enough" and keep re-irradiating someone. You might as well get it done straight away. (Senior radiographer)

The narratives above highlight areas where patient care may be hindered in the X-ray room. Radiographers commented on forcing patients into their required radiographic position (used to correctly demonstrate a fracture or pathology) on an acutely injured patient. This insight is related to the inflexible nature of some digital radiography equipment. For example, in the past X-ray film were encased in a protective cassette and remained small and untethered. Typical dimensions of X-ray film were 18×20 cm, 24×30 cm and 35×43 cm, enabling radiographers to place them in a number of positions, thus enabling a radiographer to 'work around' a patient. In contrast, and in light of 'fixed detector technology' observed at the acute hospital sites, radiographers recollect experiences whereby patients were manipulated into their radiographic position (with suspect fractures), rather than 'working around' their patient. This insight challenges the perception as to whether advancing technology does always facilitate patient experiences. Whilst the author has made some assumptions of how patients may feel (by exploring the views and attitudes of radiographers), this area of practice demonstrates an important consideration concerning the sustainable futures of patient care, highlighting that technology may not always benefit all. Further, if radiographers are 'thrown into the deep end' following 'technological pushes', requiring the operation of unfamiliar hardware, radiographers may become increasingly restricted with new equipment, which may then lead to suboptimum experiences for individuals undergoing X-ray examinations.

Commentators have identified that radiography as a profession has failed to critique or inquire into what is after all a technology-driven environment, and as a result there is inadequate consideration of radiological technology that examines its impact on the profession (Murphy 2006, p. 170). It is evident that a potential paradox exists whereby although new technologies have the ability to facilitate patient care experiences, this, in turn, remains unsustainable, especially if patient care is hindered. It is important to recognise that in more recent years, manufacturers have developed wireless digital detectors offering radiographers additional flexibility. However, not all departments have the luxury of wireless detector technology and may still utilise fixed detector equipment. In short, this form of practice may still resonate with practitioners and patients contemporarily.

As diagnostic radiography embarks into the twenty-first century, it could be argued that in order for medical imaging to ensure sustainable futures, the needs of the patient need to be assessed following purchases of new equipment. Technological push and demand pull need to be recognised and interconnected by radiology managers and critically evaluated following hardware installations. It is the social interaction between radiographer, patient and technology that is paramount in contemporary practices and, which, may also be a decisive factor highlighting whether patient care is maintained within the medical imaging profession.

12.4.2 Suboptimum Use of Ionising Radiation: Unsustainable and Unsafe Practice(s)

The clinical benefits of undergoing X-rays are generally accepted (Dalrymple-Hay et al. 2002), yet the use of ionising radiation in medicine remains the largest man-made contributor to humans. This affirms that the optimum use of ionising radiation by radiographers is essential. Before exploring the use of ionising radiation in contemporary healthcare, it is important to consider the science of radiobiology to underpin the arguments posed hereafter.

The current radiobiological paradigm conforms to a dose-response model known as 'linear non-threshold' (LNT). The LNT model is used in radiation protection today and underpins radiation risk, even at small levels. The LNT model assumes that long-term biological damage can be caused by ionising radiation (at all levels) and is directly proportional to the radiation dose received; thus, no safe level of ionising radiation exists (International Commission on Radiological Protection 2007). Whilst this model has been challenged, with some commentators suggesting that low levels of ionising radiation (used within medical imaging) may benefit patients (known as hormesis) (Allison 2009), the general acceptance resides in the theoretical LNT model. In short, no level of ionising radiation delivered within the healthcare environment is deemed 'safe'. This radiobiological paradigm currently informs legislation in the UK and is the rationale for experimental research aiming to limit and optimise ionising radiation to both patients and practitioners in the clinical environment (Department of Health 2017). The LNT model is important because it identifies that all practitioners in the UK must keep levels of ionising radiation 'as low as reasonably practicable' (ALARP)) when irradiating patients.

Radiographers remain accountable for the ionisation radiation delivered to produce a digital radiograph; thus, it is important to explore this area of practice in order to ensure the sustainability of utilising X-rays appropriately. The introduction of digital radiography was regarded as an important step at minimising radiation levels, which is supported by published literature, offering significant dose reduction whilst maintaining sound image quality (Strotzer et al. 1998a; Volk et al. 2000; Geijer et al. 2001; Bacher et al. 2003). Whilst significant levels of dose reduction have been reported (33–80%), a phenomenon known as 'dose creep' juxtaposes these dose-limiting benefits. 'Dose creep' is a widely accepted phenomenon whereby an operator of X-ray equipment can increase X-ray exposures (factors which determine quality and quantity of X-rays emitted from the X-ray tube) without effecting image quality (Seeram 2013). By increasing X-ray exposures in digital radiography, patient dose is increased but does not affect image quality.

Historically, X-ray film did not allow an operator to do this. If an operator increased levels of ionising radiation, the film would be observably 'blackened' and deemed 'overexposed'. Yet, in contemporary practices and with technological improvements in digital detectors, it is possible for practitioners to overexpose a digital detector and still acquire a sound diagnostic X-ray image. This is because a digital detector can have a dynamic range of 1:10.000 (higher than conventional X-ray film 1:30), providing a greater exposure latitude and contrast resolution. This remains a key concern for exposure error and increases to patient dose because where 'film blackening' would indicate an 'overexposure', this no longer exists in digital systems as studies suggest that the transitional process from film screen (FS) to digital could increase patient doses by up to 40–103%, with digital radiography systems autocorrecting overexposed radiographs by up to 500% (Vaño 2005; McConnell 2011). Therefore, a tenfold overexposure may not be recognised as 'too black' on a digital radiography system; thus, images are deemed 'diagnostic' by the radiographer even though the patient would have been overexposed.

These technological issues are closely linked with cultural norms. For example, radiographers may knowingly increase exposure factors in order to prevent radiological repeats, thus knowingly over-irradiating patients in order to ensure optimum image quality (Hayre 2016b). This raises two concerns. First, if radiographers are deliberately increasing exposure factors, this resonates with unsafe clinical practices, whereby the stochastic risks associated with ionising radiation are also increased to patients. Second, this 'cultural norm' impacts on the sustainable future of local X-ray environments because it may become generally accepted practice amongst peers and student radiographers, who learn by observing radiographers prior to becoming radiographers themselves. Radiographers remain at the forefront of delivering this artificial source of ionising radiation to patients globally; however, this central role remains unchallenged, unsustainable and at worst unnoticed.

A key feature of all digital radiography equipment is the exposure index (EI), which provides the practitioner 'feedback' on the appropriateness of the radiographic technique employed (Schaefer-Prokop et al. 2008; Seibert and Morin 2011). The EI provides a numerical indication of the image quality produced. Whilst the EI is not directly linked to patient dose and stochastic risk, literature suggests that the EI remains a useful tool when acquiring digital radiographs whereby it can assess if a radiographic technique selected was appropriate or not. Siebert and Morin (2011, p. 575) claim that recognition and evaluation of the EI remains paramount because it can ensure the correct use of equipment with a continuing aim of optimising radiation doses on an exam-by-exam basis. In his study of radiographers, Hayre (2016a, pp. 182-183) recorded the use and appropriateness of the EI amongst radiographers and how they utilised it to manage X-ray exposures. Some radiographers recognise the value of the EI value, often using it to reflect on their X-ray exposures delivered:

> The exposure index and that comes into play – if you did an X-ray at first and it wasn't a diagnostic image and then you say "How do I correct it?" you then go and refer to your exposure index and see what it says. And you say "OK, this is how it's going to guide me. It guides me now to make an adjustment and produce a diagnostic [image]. (Radiographer)

> Yes. So you can recognise an under-exposed film by looking at it [exposure index], but for an over-exposed film, you've got to know what ball park your exposure index should be in. (Senior radiographer)

The data above highlights that radiographers do reflect on their radiographic exposure factors by acknowledging the EI value. This supports Uffmann et al. (2008) findings whereby X-ray operators understood the EI as a control mechanism for controlling 'dose creep', thus attempting to optimise digital radiography systems. On the other hand, radiographers (from the same study) acknowledged that they do not use the EI, as some fail to locate the value, which leads to radiographers judging the radiographic image on appearance alone (Hayre 2016a, b, pp. 183–184):

> I don't, with digital radiography. Again, I'm so unfamiliar with the system, I just look at the image quality. I know what my exposure factors are, I know what the dose is, and what I expect as a dose…I don't think I'd know how to look at the exposure index on that machine. (Radiographer)

> I have not got a clue where to find the exposure index on the digital radiography system! So I can't answer that question. If it was there, if I knew where it was, then I'd be able to… No one's ever taught me that. Yes, because there's no technical way of looking at it – I just go on my own experience of what an image should look like. (Senior radiographer)

The narratives identify that image assessment by radiographers may be based on subjective knowledge, which in some respects is important to ensure sustainable radiographic practitioners and also highlights a potential danger. For example, on the one hand, solely relying on the EI value negates the role of a radiographer because if a radiographer acquires a radiograph with an exposure index indicating 'underexposure', it may remain 'diagnostically acceptable' by answering the clinical question posed by the referring clinician, a decision which can be made by a radiographer. On the other hand, radiographers failing to consult the EI may unknowingly overexpose patients, thus facilitating 'dose creep'. Because radiographers are responsible for keeping ionising radiation 'as low as reasonably practicable', it is important that radiographers recognise limitations and benefits associated with the EI in order to limit 'dose creep' wherever possible and therefore maintain autonomous practitioners. The term 'sustainability' is often associated with ongoing improvements, but it is also linked with retaining principles pertinent to a professional group (Buchanan et al. 2007). This suggests that as technology continues to evolve, with attempts to achieve 'higher standards' and better imaging performance, the medical imaging profession is required to continuously reflect and adapt to technological advances in order to maintain clinical competencies and keep doses to a minimum.

The use of general radiography continues to increase both nationally and internationally, therefore the phenomena and management of 'dose creep' should be managed locally by all radiographers. Audits, existential learning and continued professional development can help practitioners maintain excellence in this area of practice. As seen in recent years, the enhancements of image quality, underpinned by increased 'bit depth', matrix size and computing power remain hard to resist for clinical environments, yet the benefits proposed need to be critiqued to ensure that delivery of ionising radiation is sustained in future years.

12.5   Radiography Education: Existing Challenges and Opportunities for Higher Education Institutions

Exploring the sustainability of radiography education in HEIs is important to discuss. The author considers two areas for discussion. First, because radiographers not only acquire images of diagnostic quality, utilising ionising radiation to best affect, they also engage in the advanced practice of reporting radiographs, a role historically undertaken by medical doctors. This practice requires additional training and education, enabling radiographers to report a variety of medical images. Whilst there is a growing need for image interpretation by radiographers, there is some resistance to this progression, which will be discussed in Sect. 5.6.1. For now, the author focuses on a critical evaluation of both advantages and limitations to this role extension by radiographers educationally and how the central tenets of radiographer practice must be maintained prior to any role enhancements.

Second, students wishing to undertake diagnostic radiography (or other healthcare degrees allied to medicine) were previously supported financially by a bursary, supported with waiving of student tuition fees. The funding from central government arguably offered institutional security amongst HEIs, whereby degree programmes would be expected to reach maximum occupancy. Yet, following the recent cessation of central government funding, the author recognises and reflects on the current 'uncertainty' amongst staff within his HEI. In the current social-political climate, students are now required to fund health degree programmes, shifting the educational paradigm for education providers and students wishing to pursue a healthcare career. Following the removal of government funding, recent reports suggest fewer students are considering undertaking nursing degrees following the cessation of financial support (Turner 2017). This is backed up by statistics demonstrating a 23% [5% across all health subjects] reduction in applications to the Universities and Colleges Admissions Service (UCAS). In response, the author will reflect on his experiences as an educator and also offer originality by applying a 'business framework' to healthcare programmes. Few positional statements offer a healthcare business module focusing on the existing changes within the higher education sector. In short, it offers an attempt of ensuring sustainable education amongst healthcare programmes within HEIs nationally.

12.5.1   Experts of Image Acquisition or Image Interpretation? Sustaining a Sound Knowledge Base

Literature in the last two decades has focused on the role development of imaging interpretation by diagnostic radiographers. Prior to radiographers embarking on such 'advanced practices', image reporting was previously undertaken by medically educated practitioners, diagnostic radiologists, who had not only undertaken a medical degree, but later 'specialised' to become 'image interpretation experts'.

Whilst the role of the radiologist is generally accepted globally, studies in the UK continue to show that radiographers remain as competent as radiologists at undertaking image interpretation (Piper et al. 2010; Piper et al. 2014). This recognition has led to a growing number of 'reporting radiographers' nationally, who undergo postgraduate education to specialise in reporting images pertinent to their imaging modality of choice. This is now generally accepted as 'advanced practice' whereby promotion within the NHS is expected. Whilst there is a growing trend for radiographers to proceed to acquire image interpretation skills, the author reminds the reader of the central tenets of that as a diagnostic radiographer and how these should be sustained. In a recent publication, Hayre et al. (2017, p. 149) identified that radiographers may lack knowledge and understanding of image acquisition with advancing technology. He identifies that there is a potential for radiographers to be deskilled in a profession where radiographers fail to know and understand how digital radiographs are produced:

> I don't know any of the aspects behind how it [digital radiography] works. I think that's a lack of knowledge that we don't actually have [sic].(Senior radiographer)

> Up to now – I have to be honest with you – I still don't quite understand the concept [of digital radiography]. Though I can imagine what is going on I would like to know. Maybe it's not within my level, but you'd want to know the technical side of things a bit more, so that you understand the process. (Radiographer)

> Definitely. And I think I can honestly say that I don't have the necessary understanding and I am using it [digital radiography] without that knowledge. And I think it's important. (Senior radiographer)

This argument is supported by the conjecture that if radiographers do not know and/or understand image acquisition when using digital technologies, then how can they begin to optimise it? Continued improvements of image storage, 'bit depth', image matrix and crystal structure of digital hardware is expected and accepted to enhance the delivery of radiographic practice by means of reducing ionising radiation and enabling faster processing speed (Seeram 2011). However, this leads to question that if radiographers fail to understand how X-rays are captured, then this negates the primary role of radiographers nationally and internationally. At present, and as identified above, career progression tends to focus on image interpretation, with radiographers securing promotion through additional learning to become a 'reporting radiographer'. Yet, in order to ensure that radiographers do adhere to technological principles and have a sound level of knowledge and understanding, it is argued that new roles are formed clinically. For example, an 'advanced' role titled 'digital radiography champion/ambassador' could become a generally expected position for radiography departments, supported with benefits closely associated with image interpretation expertise, such as pay incentives. This role would recognise the importance of maintaining image acquisition expertise within clinical environments, which can be monitored, with regular updates to peers within the clinical environment.

Such a role could offer opportunities for dose reduction to peers, without compromising image quality. This would enhance and sustain clinical competencies by reminding and underpinning the central tenets of the role of the radiographer, which is to retain expertise of image acquisition in order to produce radiographs of diagnostic quality and keeping doses 'as low as reasonably practicable'. The importance of challenging this area of practice resides in the potential for the diagnostic radiography profession to 'shift more' towards image interpretation expertise, through promotion and recognition of 'advanced practice', something which continues to be recognised by the SoR (2017a). Whilst the author welcomes this area of specialisation (demonstrated further in Sect. 5.6.1), he merely wishes to emphasise the importance sustaining pertinent knowledge and understanding of digital technology, which should be recognised through career advancements. It is here, where HEIs and professional bodies play pivotal roles. Whilst physics, radiation protection, patient care and image interpretation remain core competencies for student radiographers, it is proposed that technology becomes mandatory within the radiography curriculum. This will entail that radiographers have pertinent knowledge and understanding of technological advances, making prospective practitioners aware of the limitations and advantages technology plays upon delivering ionising radiation. Image acquisition and image interpretation remain central at ensuring sustainable futures within the radiography profession, yet without critical recognition academically and professionally, radiographers may lose specific expertise, which remains paramount to radiological practices globally.

12.5.2 Creating Shared Value: Can This Facilitate Sustainable Futures for HEIs?

It is generally accepted that 'sustainability' for organisations will depend on the organisational setting, environment and type of work undertaken (Jacques 2015). For example, sustainability on a hospital ward, an outpatient clinic, and/or X-ray department, will differ from the sustainable goals of HEIs. However, what remains consistent is the likelihood that each organisation will be open to opportunities and constraints and presented with conditions that may either challenge or influence sustainable working practices (Galpin et al. 2015). As previously noted, a recent challenge faced by HEIs resides in the loss of financial security from Health Education England, who historically commissioned student radiographer placements (and other health programmes) to each HEI nationally. In September 2017, students enrolled onto the diagnostic radiography degree programme at the author's institution will be required to seek funding by means of self-payment and/or a student loan, in order to fund their undergraduate degree. The 23% reduction identified by UCAS for nursing and midwifery and 5% reduction for diagnostic radiography are important to recognise. Whilst these applications are expected to rise in future years, this remains conjectural, thus by offering a strategic view may help facilitate student recruitment in future years.

The field of business ethics has focused on advanced methods of how companies can engage for societal and environmental good (von Liel 2016). Corporate social responsibility (CSR) is generally accepted and a widely recognised framework utilised by large organisations in the UK, including the NHS (Scott 2010), the Metropolitan Police (Godwin 2010) and the numerous HEIs (University of London 2014; University of Manchester 2016). Yet, Porter and Kramer (2011) criticise CSR as a tokenistic gesture in attempts that organisations are now required to be 'responsible' and 'give something back to society'. In response, Porter and Kramer (2011) suggest that 'creating shared value' (CSV) offers businesses an alternate ideology enabling organisations to transform social problems into business opportunities.

The word 'customer' may not resonate wholly within the university environment, yet there is a growing acceptance that students should be perceived as customers in contemporary education (Comm and Mathaisel 2008; Consumer Rights Act 2015). Perceiving students as customers in a 'valued-based recruitment model' will arguably enable HEIs to formulate alternate pathways for prospective students. For example, some HEIs in the UK offer 'fast track' preregistration MSc courses to become registered diagnostic radiographers, a degree lasting 2 years (University of Derby 2017; Teeside University 2017). Such alternate 'products' remain 'society-focused' and may help satisfy societal needs [in particular mature students] with students only requiring to commit to 2 full years of study, instead of three. 'Value' means developing a winning strategy and is about balancing the perceived benefit of a product and the cost. If customers [students] perceive the benefit of a product as higher than the cost, then there will be a balance in the value (von Liel 2016). This remains 'valued-based' because individuals will consider a courses length of study in affiliation with other social issues. Such opportunities are currently not offered by multiple HEIs, thus to create shared value, it is the responsibility of academics to reflect and consider the individual values of potential students [customers] in order to facilitate their educational needs.

This leads the author to suggest that by CSV within this framework, HEIs should challenge the structure of educational programmes [products], prior to engaging with the student market. Porter and Kramer (2011) recognise that too many companies have lost sight of a most basic question: Is our product good enough for our customers? Or for our customers' customers? Thus one question should be asked by HEIs nationally offering healthcare education: Do we offer educational programmes that meet the needs of the social community? Whilst preregistration master programmes may help forestall attrition amongst graduates with undergraduate degrees, this does not accommodate those without a previous degree. In their article, Porter and Kramer use the example of Wells Fargo, a US bank that has developed online tools to help budget, manage credit and pay down debt, as a business that is refocusing on what is good for its customers. Similarly, HEIs can learn from such innovations and transform current curricula. Is it possible to offer full-time undergraduate students opportunities to watch/listen to lectures and seminars online? This not only limits student travel, offering students opportunities to 'attend' lectures digitally, but they also remain fully engaged in academic activities via the World Wide Web. Buchanan et al. (2007) claim that sustaining standardised working practices may be damaging to an organisation, and it is important for some initiatives and/or courses to decay, due to external circumstances. It is recognised that vocational degrees require students to attend placement settings in order to learn and demonstrate clinical competencies in a number of key areas, yet for 'academic content', it is possible that students become fully engaged online, whereby learning and reflection is plausible. In short, this offers a multifaceted approach to prospective students nationally and internationally. If academics critically reflect on alternate pedagogical/andragogical methods, utilising both face-to-face and/or virtual platforms, HEIs can innovate curriculum activity that meets the demands of society. In addition, this could create an internationally shared value by outreaching to students in other countries, which has continued to increase in the last decade (Universities UK 2015). Whilst this value-centred approach would limit time and energy costs to students, research exploring the effectiveness on students' learning should be monitored and remains paramount to HEIs, ensuring sound knowledge and understanding of academic content.

This perspective at CSV attempts to build sustainable products that continue to meet societal needs. By ensuring the longevity of students enrolling onto undergraduate and postgraduate healthcare programmes, it is felt that HEIs will need to adapt in order to grow nationally and internationally. Importantly, for HEIs to remain at the forefront of shared values, they must question and continually re-question the societal benefits of such products. Universities in the USA are beginning to incorporate social incentives by introducing 'sustainability ambassadors', 'conservation advocates' or 'eco-reps' within their education domains (Jacques 2015). These positions are arguably needed now in contemporary HEIs in the UK, whereby social and political changes continue to affect higher education decisions by numerous stakeholders.

An additional facet to 'educational products' within higher education is the notion of apprenticeships. The NHS is beginning to offer apprenticeships in nursing in 2017/18, which will require HEIs to realign marketing and strategic drivers. Whilst these are currently under development within the NHS, the NHS advocates that they are beginning to offer 'flexible routes to becoming a nurse' (NHS 2017, p. 1). In response, HEIs must ensure they similarly remain flexible to current trends within education, which after all aligns to the third largest employer globally. It is therefore central to all institutions that their underlying values are adaptable and readily transformative. This remains central to the CSV model advocated by Porter and Kramer (2011) whereby students [customers] wishing to embark on an educational programme are met with a number of products that facilitate their needs. In response, the CSV principle offers originality to challenge and support HEIs, whilst enabling them to question whether the educational products offered are 'fit for purpose' or not simply 'fit for award'. Failing to engage societal needs and adapt accordingly may lead to HEIs falling behind and losing out to competitors. In short, universities should acknowledge the requirement to engage in a possible educational paradigm shift that focuses on 'what students want' and 'how best to compete with neighbouring institutions' in order to remain educational providers and innovators.

12.6 Workplace Culture: Sustaining Radiographer Advances and Engagement in Research

In the twentieth century, 'culture' emerged conceptually in anthropology encompassing a range of human phenomena that could not be attributed to genetic inheritance (Malinowski 1923). It was first described by Sir Edward Tylor: 'culture… is that complex whole which includes knowledge, beliefs, arts, morals, law, customs and any other capabilities and habits acquires by [a human] as a member of society' (Tylor 1884). Because culture is encompassed within sustainability (as outlined by Buchanan et al. 2007), this naturally offers a range of subdivisions that can be attributed to sustainable futures. The term 'medical anthropology' was first used by Scotch (1963), describing it as a discrete field of study within medical environments. Whilst culture is predominantly accustomed to a literary or artistic heritage with prevailing values and ethos of a particular nation, hospitals have cultural significances due to their complex organisational structures and diverse specialised functions within the number of professional roles, thus pertinent when considering sustainable futures.

In the last decade, researchers have attributed 'culture' to the workplace. The way in which attitudes are expressed within a specific organisation is described as a 'corporate' or 'workplace culture'. One generally accepted and widely cited definition of workplace culture is provided by Smircich (1983, p. 339):

> Workplace cultures revolve around the shared values and attitudes and the shared experiences that validate them. A culture includes everything that is learned and shared by its members: its social heritage and rules of behaviour, its own customs and traditions, jargon and stories.

In recent years there has been increasing attention on exploring workplace cultures within the NHS following suboptimum healthcare delivery, such as those attributed with Mid Staffordshire NHS Trust (Department of Health 2013). Medical imaging departments have their own workplace culture(s) and could affect the sustainability of healthcare practices. Although workplace culture has been defined as 'the way things are done around here' (Drennan 1992, p. 3), when considering sustainable futures, this could be damaging. Buchanan et al. (2007, p. 130) claim that organisational culture plays an important role in sustainable futures following the introduction of new methodologies and/or approaches in healthcare environments. In their study they acknowledge that one of their NHS sites had a 'can-do' culture and a readiness for change, as identified by a nurse who said 'we never say no. We say we will if we can'. The readiness identified by this healthcare professional identifies an open approach to change and how it, in short, aims to benefit patient care. As identified by Smircich (1983, p. 339), workplace culture revolves around the values and attitudes of its members. The following section will now consider how adopting an anthropocentric approach can bridge opposing belief systems that remain apparent within the medical imaging profession.

12.6.1   Attitudes of Radiographer Reporting: Can Anthropocentrism Bridge Opposing Beliefs?

The view that sustainability can be hindered by individuals within a workplace culture is important to consider. This section offers readers a perspective that explores anthropocentrism in an ongoing debate by the SoR and RCR. As discussed in Sect. 5.5.1, the role of the diagnostic radiographer has evolved in the last two decades, whereby emphasis on image interpretation remains a central part to radiographic practice. The 'shift' of radiographers primarily from being image acquisition practitioners to becoming both image acquisition and image interpretation experts is generally expected in the UK and has evolved in order to sustain working practices by producing timely medical imaging reports for referring clinicians. Yet, a recent publication by the RCR juxtaposes such role development by radiographers. It affirms a critical examination [and cessation] of radiographer reporting in a recent publication titled 'The radiology crisis in Scotland: sustainable solutions are needed now' (RCR 2017). The publication argues that the role of image interpretation by radiographers is of little value in the clinical environment, claiming that 'imaging reports must be diagnostic and actionable to be of any value… and that reports produced by those [radiographers] without medical training are inevitably observational and descriptive'. The statement proposes that overseas doctors and a vast increase in radiology trainee numbers are required to fill the shortfall and to take into account the prospective rate retirements amongst radiologists. The RCR conclude by firmly discouraging radiographer (and other nondoctor reporting of medical images) and suggest that there needs to be a sustained programme of both short-term international recruitment and longer-term increases in clinical radiologist training numbers.

This view of preventing radiographers to report medical images to ensure sustainability is challenged by the SoR. Their rebuttal to the RCR statement argues that the role of the reporting radiographer is multifaceted, whereby radiographers are now integral to radiology academies, to which radiographers (reporting and otherwise) contribute to the education of trainee radiologists and have shown throughout the UK to be a competent as radiologists at detecting pathology, for both general and cross-sectional imaging (SoR 2017b)). As the professional bodies offer juxtaposed positions on 'sustainable futures' regarding image interpretation, the author considers anthropocentrism, supported with deontological and utilitarianism philosophies to help bridge opposing attitudes with the overall aim of delivering a safe and sustainable services for patients.

Regardless of the views concerning sustainable futures for image interpretation, the SoR and RCR share a common goal; ensuring patients receive optimum patient care and achieve accurate radiological diagnoses for suspected pathology. This overarching principle resonates with deontology ethics, which stresses the duty to protect and provide current and future welfares of human beings (Jacques 2015, pp. 130–131). Further, utilitarianism is an ethical theory which states that the best action is the one that maximises utility. These philosophical virtues in some way explain the evolution of radiographer reporting, whereby utilising radiographers to report medical images enabled organisations to cope with an increasing radiology workload in the diagnosis of patients with suspected healthcare conditions. Recent statistics in the UK reinforce the need for radiographer reporting. Diagnostic radiographers currently report 21% of all general radiographs undertaken in hospitals (SoR 2017a). Further, in 2015/16 the Mid Yorkshire NHS Trust utilised radiographers to report 6660 CT head examinations (ibid). It is evident that radiographer reporting is not only well-established but growing in order to sustain the needs of healthcare practices in the UK.

Whilst the views offered by the RCR are discouraging, the author questions whether this obstructive ideology is for protection of their own professional identity. Price (2001) offers insight into the origins, demise and revival of radiographer reporting. His article offers detailed historical perspectives of the radiologists struggle to be recognised within the medical curriculum [1917] and how radiologists were sometimes looked upon by physicians and surgeons as 'some sort of superior bottle washer' (Price 2001, p. 106). Interestingly, following the end of the First World War, a large number of trained 'X-ray assistants' [radiographers] were looking to establish themselves as a profession, alarming the British Medical Association. At a Medico-Political Committee, their attitudes can be gleaned from the following excerpt in 1917 (ibid, p. 107).

> While there is a mechanical side to radiography in which the lay assistant can be extremely useful to the medical radiographers it is highly undesirable that lay persons, however, skilled in this in technique, should be encouraged to set up themselves and pose as experts in the interpretation of skiagrams [radiographs]… A good many of them have acquired more self confidence in diagnosis than is good for them or for the general public.

The excerpt above not only offers an insight into limiting radiographer opportunities 21 years after the discovery of the X-ray by Röntgen in 1895; it also resonates with contemporary views advocated by the RCR. Whilst a dichotomy exists contemporarily concerning radiographer reporting, a paradigm shift is happening, whereby diagnostic radiographers are successfully undertaking roles similar to those as medical doctors. In short, this may be discerning to radiologists who are expected to undergo extensive education, via medical school, foundation year practices and specialist registrar training prior to reporting medical images, a process that could take 12–15 years. In comparison, a diagnostic radiographer could undertake some form of medical image interpretation within 4 years and then themselves educate trainee radiologist in academies throughout the UK.

By critically looking at the role development of radiographers, it could be argued that the cessation of reporting practices by radiographers allows the protecting of the radiologist's identity whilst maintaining their significance within the medical profession, as previously outlined in 1917. The danger of this resides in the sustainable futures of healthcare whereby medical imaging is becoming increasingly complex, supported with an increasing number of medical imaging examinations globally. The RCR response to 'short-term sustainability' by recruiting doctors from overseas does not consider wider sustainable issues affecting countries outside of the UK. Their long-term sustainable solution affirming to train more radiologists may take decades to reach a satisfactory equilibrium. In the meantime, by considering an anthropocentric ideology that encompasses deontology and utilitarianism, it may enable the RCR to critically look at the need for radiography reporting and accept that radiographers will continue to play a pivotal role in image reporting in years to come.

Jacques (2015) recognises that whilst a hard-nosed realist [positivist] may not consider ethics necessary to sustaining society goals, he argues that both morality and ethics remain at the centre to all sustainable commitments within healthcare. The author reminds the reader that all practitioners are legally, morally and ethically obliged to their respective regulatory bodies. Thus, in light of our current moral and ethical obligations, would the complete cessation or reduction of radiographer reporting be morally and/or ethically the right thing to do? A number of studies have found that when individuals work together towards environmental goals, it creates norms of pro-environment behaviour, whereby individuals begin to behave in more environmentally sustainable ways (Hopper and McCarl-Nielsen 1991 cited in Nordlund and Garvill 2002). Similarly, in order to maintain sustainable futures of image interpretation, professional bodies and practitioners (radiologists and radiographers alike) should be working towards a common goal of sustaining excellent outcomes for patients seeking a diagnosis of a suspected pathology.

The author suggests that sustainable futures utilising this anthropocentric philosophy could offer 'micro-sustainable approaches' within NHS hospitals, whereby members of the profession critically examine sustainable measures of image interpretation that meets the patient's needs. Such collaborations locally may begin to resonate in other organisations nationally and possibly internationally, later becoming 'macro-sustainable solutions' to image reporting and by ensuring sustainable futures of medical diagnosis. In short, utilitarianism and deontology within an anthropocentric sphere remain paramount when considering sustainable futures of the medical imaging profession. It is argued that in order to sustain clinical excellence, radiographers and radiologists need to adopt an anthropocentric model that considers ethics and the needs of patients prior to rejecting a professions ability to sustain image interpretation performance. This is important to consider because if radiographers and radiologists alike fail to challenge obstructive positional statements made by the RCR, then the practice of career advancement for radiographers in medical imaging could be open for disposal and what is generally open for disposal can be used and depleted, as seen historically (Price 2001; Jacques 2015, p. 135). In summary, diagnostic radiographers and diagnostic radiologists remain central in the patient pathway and that radiographers should continue to take on image interpretation responsibility in order to ensure the sustainability of patient outcomes.

12.6.2   Ensuring Sustainable Futures by Enhancing Our Evidence-Based Knowledge

In order to ensure that futures are sustained for healthcare delivery, there needs to be a growing emphasis on research within medical imaging profession. As discussed previously, the author challenged whether technology remained the panacea of healthcare within the radiography profession because whilst technologies work by the same universal rules, even on the moon, when applied through humanistic interaction and workplace, cultures technologies are in danger of becoming handicapped (Trompenaars and Hampden-Turner 2011). Recent research shows that advances in technology have altered ways of utilising ionising radiation, potentially deskilling radiographers and hindering patient experiences in general radiography; however, they do offer advantages that should be celebrated. Technology will continue to advance in general radiography and other imaging modalities; thus, research exploring its application clinically is important to critically reflect upon and identify areas for improvement.

The concept of sustainability is multifaceted, defined by Buchanan et al. (2007, p. 30) as 'maintaining a process of quality improvement, commitment and leadership, improving techniques, education and training, teamwork and cultural change' (Buchanan et al. (2007, p. 30). This statement resonates with all working practices in medicine, nursing and the allied health professions and playing a pivotal role in all areas of healthcare. Furthermore, because of the multifaceted elements associated with sustaining health and social care, the author offers a methodological approach that can best be utilised to enable clinical exploration and critical reflection of contemporary healthcare practices. This research methodology is ethnography. Ethnography provides a unique insight into cultures pertinent to a profession (Atkinson and Hammersley 2007). It enabled the author to critically evaluate his own profession, diagnostic radiography and explore pertinent issues, such as the use of ionising radiation, technological advances and patient care. Further, whilst primarily aligned with the qualitative paradigm, ethnography can encompass quantitative methods as part of an overall study, enhancing generalisations of the social world (Hammersley 1992). This can enable researchers to build on observations and discussions made with research participants, which can be later supported quantitatively (ibid). For example, the author's own research encompassed observations, interviews and X-ray experiments within what he termed an 'umbrella ethnographic methodology', which encompassed the use of inductive and hypothetical-deductive reasoning to capture radiographic practices (Hayre 2016a, p. 74). This is depicted in Fig. 12.1.

This innovative approach was not classified as a mixed-method approach but a postmodern methodology of undertaking healthcare research in a clinical environment. This approach is supported by Holliday (2016, p. 20), affirming that:

> My personal conclusion [to Hayre' discussion] is that there does need to be an 'umbrella' strategy of investigation or methodology within any research project that drives whatever methods of data collection and analysis are used whether they are quantitative or qualitative.

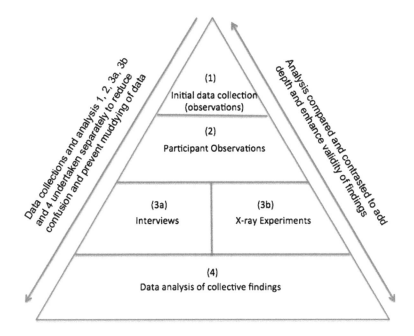

Fig. 12.1 Use of inductive and hypothetical-deductive methods in ethnography

> This means that within a postmodern paradigm the understanding that both the social
> phenomena being investigated and the methodology for investigating them are socially and
> ideologically constructed will apply also to quantitative methods.

This postmodern ethnographic methodology is an important consideration when
exploring sustainable futures within healthcare settings. Practitioners in all health
disciplines utilise some form of qualitative and quantitative reasoning in their
everyday work. Therefore, by encompassing ethnography in this postmodern
approach can offer a plethora of research opportunities for healthcare professionals.
For example, all healthcare professionals begin by observing, listening and talking
with their patients. They may begin by introducing themselves, asking their level of
pain (depending on injured state) and/or observing a wound and/or health condition.
Through utilising such 'qualitative approaches' by means of observation, discussion
and reading and/or taking notes the healthcare professional may then collect blood
cultures, request X-rays and or prescribe pain relief and/or antibiotics, each of
which has historically stemmed from research aligned to realism [positivism]. In
short, these scenarios suggest that in order to critically look at sustainable futures in
healthcare, i.e. emerging cultures, leadership, quality and improvement from a
health and social care perspective, practitioners are required to critically challenge
paradigmatic approaches in order to undertake original healthcare research. In an
ever-changing and evolving world, practitioners need to encompass a range of

attributes and rely on a number of services in order to provide and critically reflect on the care provided to patients clinically. This remains a central tenet of ensuring sustainable practices for not only medical imaging but health and social care in general.

12.7 Conclusion

This chapter began with an examination of the term 'sustainability' for the medical imaging profession and how it remains pertinent to continuously examine and re-examine local, national and international topics central to the profession. The multifaceted acceptance of the term suggests that 'sustainable futures' encompasses an array of practice elements that impact on the professions, diagnostic radiography and diagnostic radiology alike and also how it may impact on the healthcare delivery of patients. In the past, sustainability has primarily focused on limiting ionising radiation and prospective workforce concerns, yet this chapter demonstrates wider sociocultural and political issues that remain pertinent to the sustainable futures of the medical imaging profession. Three areas have been discussed.

First, whilst it is general acceptance that technological advancements have relieved tensions following the 'digital radiograph', the transformation from X-ray film to a-Se fixed detector technology may offer hindrances towards delivery of patient care. For example, original data offers readers insight into the potential discomfort to patients upon undergoing radiographic examinations. Further, the widely accepted phenomena 'dose creep' may be uncontrolled in the clinical environment as some practitioners fail to locate and/or reflect upon the exposure index post-radiological exposure. By challenging the sustainability of ionising radiation usage amongst practitioners, it highlights that patients may continue to receive higher radiation doses than necessary if it remains commonplace. This, in short, must be recognised in order to ensure that ionising radiation is kept 'as low as reasonably practicable' to limit any occurrence of stochastic affects.

Second, the author discusses two themes that are important to the diagnostic radiography curriculum. It is affirmed that radiographers should embark on additional expertise in the form of image interpretation, yet the author argues that knowledge and understanding of acquiring radiographic images remain paramount if radiographers are to retain image acquisition expertise in clinical environments. Further, following cessation of tuition fees to prospective healthcare professionals, the author uses a business ideology, 'CSV', in order for academics to critically reflect on academic programmes [products] currently offered and how they may or may not be suitable for prospective students [customers]. It is argued that HEIs will need to be continuously reflexive on their curriculum delivery and question whether it meets societal goals. It is proposed that a 'sustainable ambassadors' could be utilised to foresee and discuss up and coming challenges and opportunities pertinent to the higher education sector.

Third, culture and values are argued to be important if the futures of medical imaging are to be sustained. The RCR calls for the cessation of radiographer reporting, a practice demonstrated to be successful in the last two decades. Whilst this could be attributed to protecting the professions own identified, as identified in 1917 following the rise of 'X-ray assistants' [radiographers], the authors calls for recognition that radiographers play pivotal roles in contemporary healthcare delivery and the education of radiology trainees. It is proposed that by considering anthropocentrism as a framework that encompasses deontology and utilitarianism principles, the professional bodies share common moral and ethical goal, the delivery of sound medical diagnosis and care for patients.

What remains apparent is the multifaceted nature of the term sustainable futures within health and social care. The NHS remains the third largest organisation in the world where medical imaging practices remain paramount to diagnosis in order for effective treatment and management. The topics discussed within this chapter provide insight into a profession that is wide ranging and complex. Clearly, social, cultural and political drivers will continue to impact on the professions, yet by critically engaging with wide 'sustainable futures', it enables ongoing discussions and debate within the medical imaging, which in short, serves patients in their care.

References

Allison, W. (2009). *Radiation and reason – The impact on a culture of fear*. York: Wade Allison Publishing.

Atkinson, P., & Hammersley, M. (2007). *Ethnography – Principles in practice* (3rd ed.). New York: Routledge.

Bacher, K., Smeets, P., Bonnarens, K., De Hauwere, A., Verstraete, K., & Thierens, H. (2003). Dose reduction in patients undergoing chest imaging: Digital amorphous silicon flat-panel detector radiography versus conventional film-screen radiography and phosphor-based computed radiography. *American Journal of Roentgenology, 181*(4), 923–929.

Barley, S. R. (1986). Technology as an occasion for structuring: Evidence from observations of CT scanners and the social order of radiology departments. *Administrative Science Quarterly, 31*(1), 78–108.

Becker, H. S., Geer, B., Hughes, E. C., & Strauss, A. L. (1961). *Boys in white – Student culture in medical school*. New Brunswick: Transaction Publishers.

Blau, P. M., Falbe, C. M., McKinley, W., & Tracy, P. K. (1976). Technology and organization in manufacturing. *Administrative Science Quarterly, 21*(1), 20–40.

Buchanan, D. A., Fitzgerald, L., & Ketley, D. (2007). *The sustainability and spread of organizational change*. London: Routledge.

Comm, C. L., & Mathaisel, D. (2008). Sustaining higher education using Wal-Mart's best supply chain management practices. *International Journal of Sustainability in Higher Education, 9*(2), 183–189.

Consumer Rights Act. (2015). Consumer Rights Act 2015. [Online] Available at: http://www.legislation.gov.uk/ukpga/2015/15/pdfs/ukpga_20150015_en.pdf. Accessed 12 Dec 2016.

Dalrymple-Hay, M. J., Rome, P. D., Kennedy, C., & McCaughan, B. C. (2002). Pulmonary metastatic melanoma – The survival benefit associated with positron emission tomography scanning. *European Journal of Cardiothoracic Surgery, 21*(4), 611–614.

Department of Health. (2013). Report of the Mid Staffordshire NHS Foundation Trust public inquiry. [Online] Available at: https://www.gov.uk/government/publications/report-of-the-mid-staffordshire-nhs-foundation-trust-public-inquiry. Accessed 30 Aug 2017.

Department of Health. (2015). The NHS Constitution for England. [Online] Available at: https://www.gov.uk/government/publications/the-nhs-constitution-for-england. Accessed 30 Aug 2017.

Department of Health. (2017). Ionising Radiation (Medical Exposure) Regulations 2000 (IRMER). [Online] Available at: https://www.gov.uk/government/publications/the-ionising-radiation-medical-exposure-regulations-2000. Accessed 30 Aug 2017.

Drennan, D. (1992). *Transforming company culture*. London: Mcgraw-Hill.

Fett, M. (2000). *Technology, health and health care*. Canberra: Commonwealth Department of Health and Aged Care.

Galpin, T., Whitttington, J. L., & Bell, G. (2015). Is your sustainability strategy sustainable? Creating a culture of sustainability. *Corporate Governance: The International Journal of Business in Society, 15*(1), 1–17.

Garetti, M., & Taisch, M. (2012). Sustainability manufacturing: Trends and research challenges. *Production Planning & Control, 23*(2–3), 83–104.

Geijer, H., Beckman, K. W., Andersson, T., & Persliden, J. (2001). Image quality vs. radiation dose for a flat panel amorphous silicon detector: A phantom study. *European Journal of Radiology, 11*(9), 1704–1709.

Godwin, T. (2010). Corporate social responsibility strategy 2010–2013. [Online] Available at: http://www.met.police.uk/about/documents/csr_strategy.pdf. Accessed 18 Jan 2017.

Hammersley, M. (1992). *What's wrong with ethnography*. New York: Routledge.

Hayre, C. M. (2016a). Radiography observed: An ethnographic study exploring contemporary radiographic practice. Ph.D. thesis, Canterbury Christ Church University.

Hayre, C. M. (2016b). 'Cranking up', 'whacking up' and 'bumping up': X-ray exposures in contemporary radiographic practice. *Radiography, 22*(2), 194–198.

Hayre, C. M., Eyden, A., Blackman, S., & Carlton, K. (2017). Image acquisition in general radiography: The utilisation of DDR. *Radiography, 23*(2), 147–152.

Health Protection Agency. (2011). Scale of UK exposure to X-rays revealed. [Online] Available at: http://www.hpa.org.uk. Accessed 06 May 2011.

Holliday, A. (2016). *Doing and writing qualitative research* (3rd ed.). London: Sage.

Hopper, J. R., & McCarl-Nielsen, J. (1991). Recycling as altruistic behaviour. Normative and behavioural strategies to expand participation in a community recycling program. *Environment & Behaviour, 23*, 195–220.

International Commission on Radiological Protection. (2007). *Annuals of the ICRP*. Exeter: Elsevier.

Jacques, P. (2015). *Sustainability: The basics*. London: Routledge.

Malinowski, B. (1923). The problem of meaning in primitive languages. In C. K. Ogden & I. A. Richards (Eds.), *The meaning of meaning* (pp. 146–152). London: Routledge & Kegan Paul.

McConnell, J. (2011). *Index of medical imaging*. Chichester: Wiley-Blackwell.

Murphy, F. J. (2006). The paradox of imaging technology: A review of the literature. *Radiography, 12*(2), 169–174.

National Health Service. (2017). Studying nursing. [Online]. Available at: https://www.healthcareers.nhs.uk/I-am/considering-or-university/studying-nursing. Accessed 18 Jan 2017.

Nemet, G. F. (2009). Demand-pull, technology-push, and government-led incentives for non-incremental technical change. *Research Policy, 38*(5), 700–709.

Nordlund, A. M., & Garvill, J. (2002). Value structures behind roenvironmental behaviour. *Environment and Behaviour, 34*(6), 740–756.

Philips B. (2011). Create a premium DR room like no other. [Online] Available at: http://www.healthcare.philips.com/us_en/products/xray/products/radiography/digital/digitaldiagnost/. Accessed 30 Aug 2017.

Picano, E. (2004). Sustainability in medical imaging. *British Medical Journal, 328*(7439), 578–580.

Piper, K., Buscall, K., & Thomas, N. (2010). MRI reporting by radiographers: Findings of an accredited postgraduate programme. *Radiography, 16*, 136–142.

Piper, K., Cox, S., Paterson, A., et al. (2014). Chest reporting by radiographers: Findings of an accredited postgraduate programme. *Radiography, 20*, 94–99.

Porter, M. E., & Kramer, M. R. (2011). The big idea: Creating shared value. Harvard Business Review. January–February.

Price, R. C. (2001). Radiographer reporting: Origins, demise and revival of plain film reporting. *Radiography, 7*(1), 105–117.

Royal College of Radiologists. (2015a). Sustainable future for diagnostic radiology: Working for alternative and/or multiple providers. [Online] Available at: https://www.rcr.ac.uk/publication/sustainable-future-diagnostic-radiology-working-alternative-andor-multiple-providers. Accessed 30 Aug 2017.

Royal College of Radiologists. (2015b). Sustainable future for diagnostic radiology: Flexible home working. [Online] Available at: https://www.rcr.ac.uk/publication/sustainable-future-diagnostic-radiology-flexible-home-working. Accessed 30 Aug 2017.

Royal College of Radiologists. (2016). Sustainable future for diagnostic radiology: Lifelong learning: Delivering education and training for a sustainable workforce. [Online] Available at: https://www.rcr.ac.uk/system/files/publication/field_publication_files/bfcr161_sustainable_learning.pdf. Accessed 30 Aug 2017.

Royal College of Radiologists. (2017). The radiology crisis in Scotland: Sustainable solutions are needed now. [Online]. Available at: https://www.rcr.ac.uk/posts/radiology-crisis-scotland-sustainable-solutions-are-needed-now. Accessed 30 Aug 2017.

Schaefer-Prokop, C., Neitzel, U., Venema, H. W., Uffmann, M., & Prokop, M. (2008). Digital chest radiography: An update on modern technology, dose containment and control of image quality. *European Radiology, 18*(9), 1818–1830.

Scott, J. (2010) Royal United Hospital Bath NHS Trust. [Online] Available at: http://www.ruh.nhs.uk/about/annual_report/documents/social_responsibility_report_2009-10.pdf. Accessed 31 Jan 2018.

Scotch, N. A. (1963). Medical anthropology. *Biennial Review of Anthropology, 3*, 30–68.

Seeram, E. (2011). *Digital radiography: An introduction for technologists*. Burnaby, BC: British Columbia Institute of Technology.

Seeram, E. (2013). Radiation dose optimization research: Exposure technique approaches in CR imaging – A literature review. *Radiography, 19*(4), 331–338.

Seibert, J. A., & Morin, R. L. (2011). The standardized exposure index for digital radiography: An opportunity for radiation dose to the pediatric population. *Pediatric Radiology, 41*(5), 573–581.

Smircich, L. (1983). Concepts of culture and organisational analysis. *Administrative Science Quarterly, 28*(3), 339–358.

Society of Radiographers. (2016). Reforming healthcare education funding: Creating a sustainable future workforce. Available at: https://www.sor.org/sites/default/files/scor_response_to_reforming_healthcare_education_funding_consultation_1.pdf. Accessed 30 Aug 2017.

Society of Radiographers. (2017a). Advanced practitioner. [Online] Available at: https://www.sor.org/learning/document-library/education-and-professional-development-strategy-new-directions/advanced-practitioner. Accessed 30 Aug 2017.

Society of Radiographers. (2017b). Radiographers need to be further educated enhance roles. [Online] Available at: https://www.sor.org/news/radiographers-need-be-further-educated-enhance-roles. Accessed 30 Aug 2017.

Strotzer, M., Gmeinwieser, J. K., Volk, M., Frund, R., Seitz, J., & Feuerbach, S. (1998). Detection of simulated chest lesions with reduced radiation dose: Comparison of conventional screen-film radiography and a flat-panel X-ray detector based on amorphous silicon (a-Si). *Investigative Radiology, 33*(2), 98–103.

Teeside University. (2017). PgDip/MSc diagnostic radiography (pre-registration). [Online] Available at: http://www.tees.ac.uk/Postgraduate_courses/Health_&_Social_Care/PgDip_MSc_Diagnostic_Radiography_(pre-registration).cfm.

Trompenaars, F., & Hampden-Turner, C. (2011). *Riding the waves of culture: Understanding cultural diversity in business* (2nd ed.). London: Nicholas Brealey Publishing.

Turner, C. (2017/February). Students turn back on nursing degrees in wake of Government decision to axe NHS bursaries, official figures suggest. *The telegraph*. [Online] Available at: http://www.telegraph.co.uk/education/2017/02/02/students-turn-backs-nursing-degrees-wake-government-decision/. Accessed 30 Aug 2017.

Tylor, E. B. (1884). American aspects of anthropology. *Popular Science Monthly, 26*, 152.168.

Uffmann, M., Schaefer-Prokop, C., & Neitzel, U. (2008). Balance of required dose and image quality in digital radiography. *Der Radiologe, 48*(3), 249–257.

University of Derby. (2017). MSc diagnostic radiography (pre-registration). [Online] Available at: http://www.derby.ac.uk/courses/postgraduate/diagnostic-radiography-msc/. Accessed 18 Jan 2017.

University of London. (2014). Corporate social responsibility. [Online] Available at: http://www.london.ac.uk/5634.html. Accessed 01 Jan 2017.

University of Manchester. (2016). Social responsibility. [Online] Available at: http://www.manchester.ac.uk/discover/social-responsibility/. Accessed 18 Jan 2017.

Universities UK (2015). Patterns and trends in UK higher education. [Online] Available at: http://www.universitiesuk.ac.uk/policy-and-analysis/reports/Documents/2015/patterns-and-trends-2015.pdf. Accessed: 01 Dec 2016.

Vaño, E. (2005). ICRP recommendations on managing patient dose in digital radiology. *Radiation Protection Dosimetry, 114*(1–3), 126–130.

Volk, M., Strotzer, M., Holzknecht, N., Manke, C., Lenhart, M., Gmeinwieser, J., Link, J., Reiser, M., & Feuerback, S. (2000). Digital radiography of the skeleton using a large-area detector based on amorphous silicon technology: Image quality and potential for dose reduction in comparison with screen-film radiography. *Clinical Radiology, 55*(8), 615–621.

von Liel, B. (2016). *Creating shared value as future factor of competition*. Springer: Germany.

Chapter 13
'Safety and Cybersecurity in a Digital Age'

Simon Dukes

In October 2014, at an event in Canary Wharf in London, Martin Smith MBE, Chairman of the protective security consultancy, *The Security Company,* said the following: 'if I throw a brick through a jewellery shop window and steal a watch, it is not a "brick-crime." If we insist on calling significant and increasing instances of fraud, financial scams and illegal pornography "cyber-crimes", we risk obfuscation by focusing the eye not on the crime but on the manner of its conveyance'.

It is hard not to have some sympathy for Smith's view. By adding the prefix 'cyber', we link the resulting word with computers and the digital and online world – it has been an increasing trend and is often used to make something seem contemporary and fresh. However, information technologists and information security professionals are now split on whether the term 'cyber' has been a blessing or a curse.

If a Board of a large FTSE or Dow Jones Company is made aware of 'cybersecurity weaknesses' within its organisation, then you will have their attention – and possibly their agreement to spend money to explore and mitigate those vulnerabilities. The importance of plugging any gaps in cyber defences is high particularly amongst large corporates, and in the USA there is a legal requirement to report data losses and breaches – as there will be in the UK from May 2018.[1]

But talk about cybersecurity to small- and even medium-sized organisations in Europe or the USA, or indeed to the citizen, and a fog of confusion descends: what does it mean? Does it really affect me/my organisation? Where do I go for impartial and independent advice? And how much will it cost?

Whether you are for or against the 'cyber' prefix, it is Smith's observation that is apposite. It is the word to which the cyber prefix is attached which should be our

---

[1] https://ico.org.uk/for-organisations/data-protection-reform/overview-of-the-gdpr/breach-notification/

S. Dukes (✉)
UK's Fraud Prevention Service, London, UK
e-mail: Simon.Dukes@cifas.org.uk

© Springer International Publishing AG, part of Springer Nature 2018   241
M. Dastbaz et al. (eds.), *Smart Futures, Challenges of Urbanisation, and Social Sustainability*, https://doi.org/10.1007/978-3-319-74549-7_13

focus. After all, a cyber*café* still sells coffee alongside internet access; around the world, many cyber-*romances* have led to a physical relationship; no matter what the context, cyber*bullying* and cyber*stalking* are harassment; and cyber*security* is the process by which you can protect yourself, your family, your company or indeed your country against the challenges, issues and threats we have always faced. But as this chapter will demonstrate, cybersecurity is as much about the security surrounding people and physical infrastructure as it is about the online and digital environment.

13.1   What Is Cybercrime?

The short answer to this question is that it is an offence, punishable by law, which has an online or digital dimension. In the world of modern law enforcement, cyber-crime tends to be a catch-all term for two different types of activity.

Cyber-Enabled Crimes[2]   These are offences which in a non-cyber-prefixed form have been perpetrated for centuries but can also be carried out online. Examples include all types of fraud (as we will see later, by far the most widespread and most concerning cyber-enabled crime in the UK), piracy and forgery (e.g. through illegal file sharing sites), buying and selling of illegal items (not just on the so-called dark web but also through mainstream sites), malicious communications (e.g. bullying, 'trolling' and harassment via social media), publishing private sexual imagery without consent (so-called revenge pornography), all manner of sexual offences against children from imagery to grooming and extreme pornography.

Cyber-enabled crimes are widespread and can be simply committed – the December 2016 example of a shop worker at the retail store, Halfords, stealing goods in order to subsequently sell them using a false identity on Facebook typifies the way cyber enablement can provide a method to sell on stolen goods with greater ease and anonymity than would have been the case in the non-cyber world.

Cyber-Dependent Crimes[3]   These are types of illegal activity that can *only* be carried out in an online environment, where a computer or IT system is used to attack another computer or IT system. Examples include creating malicious software for financial gain (creating viruses and worms or so-called ransomware which can be used to extort money from a victim in return for releasing control of a computer or system), illegal intrusion on to a computer system or network for the purposes of accessing/copying, vandalising or altering data stored there (hacking), or the deliberate interference or disruption of the operational effectiveness of a digital system (often termed as DDoS or distributed denial-of-service attacks).

---

[2] http://www.cps.gov.uk/legal/a_to_c/cybercrime/#a08

[3] http://www.cps.gov.uk/legal/a_to_c/cybercrime/#a08

Although ransomware is gaining in notoriety – especially after the NHS was affected by the 'WannaCry' virus in the first half of 2017 – it is the distributed denial-of-service attack that is the most apparent cyber-dependent crime for most citizens. In early 2017, there were a series of DDoS attacks on UK banks – particularly Barclays and Lloyds Banking Group. While no money was stolen, the DDoS against Lloyds Bank was prolonged. Most DDoS attacks last for a matter of hours, but the Lloyds attack started on the morning of 11 January and lasted until 13 January – thus preventing many customers from accessing the banks' online services.

However, whether cyber dependent or cyber enabled, as can be seen from the Halfords and Lloyds Bank examples, there are two characteristics that all cybercrimes have in common and which make them very appealing to a criminal or an attacker:

Scalability With the world at their fingertips via the World Wide Web, it is as resource efficient for a criminal to carry out one crime or attack as it is to carry out many, at speed.

Anonymity The perpetrator does not have to come in to physical contact with the victim; indeed neither attacker nor victim need to leave their own homes or offices for a crime to take place. And in some instances, a victim may not even know that an attack has taken place at all.

There is also one other attribute of cybercrime that make it increasingly the crime of choice for today's criminals – the chances of getting caught are incredibly small. Moreover, if the perpetrator is brought before the criminal justice system in the UK, it is unlikely that any punishment will be as severe as that for other crimes providing the same 'return on investment'.

13.2   From the Physical to the Digital

In the 1980s, committing an identity crime (stealing somebody's identity in order to purchase goods or apply for credit in their name, or to take over their existing bank and credit accounts and use them for a similar purpose) was a specialist activity which required knowledge and contacts within the criminal 'underworld'. There was risk involved, and it was often time-consuming and occasionally ... pungent.

This was because it often began with trip to the victim's house to rummage through their refuse in order to find enough personal information to apply for a duplicate birth certificate. Once obtained, a document of identity needed to be procured – either a fraudulently obtained genuine passport or a forgery from an illegal passport 'factory' run by criminal gangs. Then a trip to a bank was required with enough supporting documentation to apply for an account or credit. At any stage, there was a risk of being caught out or foiled, in which case the process would have to start again with a fresh victim.

However, if successful, the criminal would often use the same identity and documents to apply for accounts at other banks and perhaps a number of store cards as well. If you were the victim of an identity crime in the 1980s, you would have been a multiple victim because your identity would be used again and again in order for the criminal to get maximum value for the time and effort expended to secure the necessary documentation and take the necessary risk.

Today, identity crime is an industrial process with little expert knowledge required. There is enough personal information online to build-up a victim's identity quickly and easily: the huge numbers of data breaches add to the stock of identity-related details available for purchase on the dark web; and we seem to be ever more willing to share our most personal details on social media. Moreover, illegal criminal marketplaces also sell instructions and guidance on how to carry out identity (as well as many other cyber-enabled and cyber-dependent) crimes.

Once obtained, the stolen identities (because why would you limit yourself to one when so many are available for the same effort?) can be used to apply for online shopping accounts, bank loans and credit card accounts – the online application processes have in-built instant decisioning providing the criminal with immediate feedback on success or failure of the crime. With so many identities to choose from, it is likely that today, if you are the victim of an identity crime, your identity will be used once and then disposed of. This reduces the criminals' risk of overexposure on a particular identity and makes it even less likely that he (and it is usually a male that commits this type of crime) will get caught.[4]

Interestingly, what has remained pretty much as a constant from 1980 to the present day is the penalty in the unlikely event of getting caught. For an identity-related fraud, the maximum sentence in the UK is seven years imprisonment. Given the potential financial gains from this type of crime, it is little wonder that many criminals have moved to cyber fraud from other more high-risk criminal activity.

13.3   Who Is Carrying out Cyberattacks?

As implied from the identity crime example above, anybody can get involved in cybercrime. No specialist knowledge is required as the online criminal marketplace provides all the tools, techniques and tips required to carry out a range of cyber-enabled and cyber-dependent activity. Today's cyber criminals or 'actors' as they are sometimes known tend to fall in to one of six groups. However, there is always a danger in categorising in this way – because inevitably, there are grey areas, blurring around the edges of the definitions and doubling up of roles.

---

[4] https://www.cifas.org.uk/insight/fraud-risk-focus-blog/new-cifas-data-reveals-record-levels-identity-fraud

Organised (criminal) groups Carrying out large-scale cyber-enabled financial crimes, and/or creating malicious software for their own criminal use or for sale on to others and/or hacking in to systems for financial gain.

Hacktivists Often highly skilled individuals or groups using their coding and IT abilities often for protest but also for disruption and impact.

Loners Those sometimes lower-skilled criminals utilising the cyber tools created by others for financial gain or individuals involved in cyber-enabled sexual exploitation of children.

Terrorists Principally using the Internet as a propaganda tool for their cause or causes: recruiting others and spreading fear.

Nation States Using the full range of online techniques to steal data, cause disruption, carry-out espionage or reconnaissance and enable propaganda.

Opportunists Often individuals who find themselves in a situation which they realise they can exploit for criminal gain: employees of a company with access to valuable data, for example.

The difficulty in attributing cybercrimes comes from the anonymity aspect mentioned earlier – there are many ways the perpetrators can disguise their origins – but there can also be an interconnectedness between the above categories. Perhaps unsurprisingly, there are reports of nation states using criminal groups to carry out activities on their behalf in order to create deniability and avoid diplomatic complexities. Similarly there are instances of criminal groups recruiting or encouraging opportunists to obtain access to an internal system; and the skills of some hacktivists are in demand by criminals and nations (and increasingly companies: both the unscrupulous and the ethical penetration testers) for their ability to find and exploit vulnerabilities in IT infrastructure.

13.4   How Does Cybercrime Happen?

It is perhaps too obvious to state that for a cybercrime to take place, there has to be an attacker (or attackers) and a victim (or victims). But it is important to understand that there has to be a person involved at both ends – computers, currently at least, will only do what a human tells them.

And it is the human that creates the opportunity for the criminal or attacker, whether for cyber-enabled or cyber-dependent criminality. A poorly designed and secured network will allow a hacker to gain entry undetected; inadequate or out of date protect/detect software will allow malicious software to infect a system; staff members ignorant of what a potentially dangerous email might look like, or unaware that they should not click on links or open attachments in emails that they are unsure of can compromise a company's network; and individuals divulging their passwords and PINs or using public Wi-Fi to conduct their banking risk financial loss through the take-over of their accounts by criminals.

Just as there is an increasingly large cybersecurity industry providing technical assistance, advice, guidance and training to businesses in how to keep cybersecure, and just as the UK Government and law enforcement have spent time, effort and resource in raising awareness of cybercrime to its citizens, so the attackers have innovated and evolved their techniques to create new ways of carrying out criminality online – often mimicking the technology that legitimate organisations use to interact with their customers.

The cyber arms race has spawned a lexicon of terms – each posing its own unique threat to organisations and individuals:

Hacking[5] A very commonly used term to describe the unauthorised access to or control over a computer network or system for an illicit purpose.

Phishing[6] An email intended to trick you into giving out personal information such as your bank account numbers, passwords and credit card numbers but also used to inject malicious software onto a system.

Spear-Phishing[7] An email designed as above but which, from the victim's perspective, appears to be from an individual or business that is known to them – hence increasing the chances of success for the criminal.

Whaling[8] Phishing or spear-phishing attacks directed specifically at senior executives and other high-profile targets within businesses.

Smishing[9] Using SMS mobile phone text messages to induce people to divulge their personal information.

Social Engineering[10] The use of deception to manipulate individuals into divulging confidential or personal information that can then be used for fraudulent purposes – sometimes known as 'human hacking'.

Ad Clicking[11] The use by a criminal of a victim's computer to click a specific link and in doing so to increase the revenue from pay-per-click advertising.

---

[5] https://economictimes.indiatimes.com/definition/hacking

[6] https://www.microsoft.com/en-us/safety/online-privacy/phishing-symptoms.aspx

[7] https://www.kaspersky.co.uk/resource-center/definitions/spear-phishing

[8] https://digitalguardian.com/blog/what-whaling-attack-defining-and-identifying-whaling-attacks

[9] https://us.norton.com/internetsecurity-emerging-threats-what-is-smishing.html

[10] https://us.norton.com/internetsecurity-emerging-threats-what-is-social-engineering.html

[11] https://www.spamlaws.com/how-ppc-fraud-works.html

Vishing[12] Using social engineering over the telephone to gain access to private personal and financial information from the citizen – usually for fraudulent purposes.

Man-in-the-Middle Attack[13] When a perpetrator gets in the middle of a communication to eavesdrop or impersonate: typically mimicking a log in screen for an account.

Webcam Managing[14] When a criminal takes over a webcam often used to capture images later used for blackmailing a victim.

Ransomware[15] A type of malicious software that prevents or limits users from accessing their system, either by locking the system's screen or by locking the users' files until a ransom is paid.

Keylogging[16] The use of malicious software on a computer to record every keystroke made by the user, especially in order to gain fraudulent access to passwords.

Tabnapping[17] A method of changing an open tab on a computer screen to resemble a familiar log in to dupe the user in to revealing username and password. This scam typically targets people who keep multiple tabs of logged in accounts open in their browser.

And just as we should be wary of categorising the people carrying out cybercrime, so we need to be careful about using these terms too: for the citizen and small business owner, adding to the mystery of the cyber-linked crime by using 'geek speak' is unhelpful if we want to engage them in prevention and security; and unsurprisingly attackers will not use just one technique in order to get what they want but use their entire armoury (cyber and non-cyber). Focusing on one aspect of cyber-criminality may leave a victim open to other techniques.

A good example of this is the recent rise of *CEO Fraud*. This is when a criminal pretends to be a senior person in an organisation and persuades a more junior member of staff to transfer funds or provide sensitive data to the criminal. It can involve the use of social engineering techniques, vishing and spear-phishing emails.

---

[12] https://www.actionfraud.police.uk/fraud-az-vishing

[13] https://www.computerhope.com/jargon/m/mitma.htm

[14] http://www.nationalcrimeagency.gov.uk/news/960-help-available-for-webcam-blackmail-victims-don-t-panic-and-don-t-pay

[15] http://www.bbc.co.uk/news/av/technology-35091536/technology-explained-what-is-ransomware

[16] https://www.creditrepair.com/articles/identity-theft/what-are-keyloggers-and-how-do-they-steal-your-identity

[17] http://hackersonlineclub.com/tab-napping/

According to the City of London Police,[18] in 2016, an unnamed global healthcare company lost a substantial sum of money when a man who purported to be a senior member of overseas staff, telephoned a female Financial Controller who was based in one of the company's Scottish offices and asked her to transfer money to accounts in Hong Kong, China and Tunisia. The Financial Controller exchanged several calls with him as well as emails and was persuaded into transferring money into three foreign bank accounts which resulted in the company losing £18.5 million.

13.5   What Are the 'Proceeds' of Cyberattack?

Aside from some hacktivists, terrorists and those involved in child sexual exploitation or extreme pornography, the potential financial benefit from cybercrime is easy to see: applying for a loan or credit in somebody else's identity, accessing or taking over a bank account or its funds, or demanding a ransom payment provides an immediate financial return on investment. However for nation states and some criminal groups, data is the real prize.

If 'data is the new oil',[19] then we are all potentially victims of cybercrime. In July 2017, investigative journalists from the Guardian[20] discovered personal data from Australia's Medicare Card system being bought and sold online by criminal groups who were using them to create identities for fraud. The details of each Medicare identity was being traded for around AUS$30 each.

Cybercriminals will harvest snippets of personal data: mobile and landline telephone numbers, addresses, bank account and credit card numbers and join them together to create useable identities to exploit directly or sell them on to other criminals and groups to do so.

Official Russian and Chinese agencies and nation state-backed groups have stolen data in order to identify people in sensitive government roles or to help them build-up a picture of a corporate or government organisational structure for use in the future or to help influence covertly political activity in another country. And in January 2018, CIA Director Mike Pompeo told the BBC that he had 'every expectation' that Russia would attempt to influence the US midterm elections later that year.

In the run-up to the UK General Election in 2017, the National Cyber Security Centre[21] warned all political parties their election campaigns could be hacked by Russian cyber-agencies trying to influence the outcome of the election. The NCSC also warned of other state sponsors of cyberattack activity potentially hostile to the

---

[18] https://www.actionfraud.police.uk/news/action-fraud-warning-after-serious-rise-in-ceo-fraud-feb16

[19] https://www.economist.com/news/leaders/21721656-data-economy-demands-new-approach-antitrust-rules-worlds-most-valuable-resource

[20] https://www.theguardian.com/australia-news/2017/jul/04/the-medicare-machine-patient-details-of-any-australian-for-sale-on-darknet

[21] https://www.ncsc.gov.uk/news/statement-guidance-political-parties-and-their-staff

UK like Iran, North Korea and China launching attacks on or hacking in to emails, strategy documents and personal computers.

For Chinese state-owned enterprises and some unscrupulous companies, that data also includes intellectual property as well as sensitive corporate strategic information on future direction and markets, mergers and acquisitions. In early 2017 Chinese hackers were accused of building a vulnerability into software in an attempt to spy on the UK's top businesses. It was spread through a compromised update for server management software from tech firm *NetSarang*. A statement from *NetSarang* at the time said: 'Regretfully, the build release of our full line of products on 18 July 2017 was unknowingly shipped with a back door, which had the potential to be exploited by its creator'.[22]

13.6 What Is the Scale of Cybercrime in the UK?

From a citizen perspective, the 2017[23] Crime Survey for England and Wales (CSEW) compiled by the Office for National Statistics (ONS) showed there were an estimated total of 11 million incidents of crime covered by the survey for the 12 months ending March 2017. Of those, 3.4 million incidents were fraud – with over half being 'cyber related' – mainly 'bank and credit account' fraud followed by 'non-investment' fraud, such as fraud related to online shopping or fraudulent computer service calls.

In addition, according to the ONS, respondents experienced an estimated 1.8 million 'computer misuse incidents' – around two-thirds of which were 'computer virus related', and around one-third were related to 'unauthorised access to personal information' (hacking). The remaining six million incidents of crime covered by the survey constituted traditional thefts, criminal damage, robbery and violent crime, and these saw a seven percent reduction compared with the previous year's survey.

However, the fraud and cybercrime figures are 'Experimental Statistics' from the ONS – which means that 2017 was the first year for a full 12 months data capture and there are no figures from previous years to help determine trends. Yet statistics from Cifas,[24] the UK's fraud prevention service, for the first half of 2017, indicated a further rise in certain types of online fraud – particularly identity fraud. Cifas figures underline that identity fraud is now one of the most prevalent frauds in the UK and has grown steadily since Cifas' formation in 1988. During the first six months of 2017, around 500 identities were been stolen every day, and the indications from Cifas were that this particular crime would continue its rise. Cifas also underlined that over 80% of identity fraud was cyber-enabled.

---

[22] http://www.businesswire.com/news/home/20170815006022/en/ShadowPad-Attackers-Hid-Backdoor-Software-Hundreds-Large

[23] https://www.ons.gov.uk/peoplepopulationandcommunity/crimeandjustice/bulletins/crimeinengland andwales/yearendingmar2017#main-points

[24] https://www.cifas.org.uk/newsroom/identity-fraud-soars-to-new-levels

In the corporate world, according to the Cyber Security Breaches Survey for April 2017[25] (a joint report by the UK Department of Culture, Media and Sport, Ipsos MORI and the University of Portsmouth), nearly half of all UK businesses suffered a cyber breach or attack in the past 12 months.

The survey revealed nearly seven in ten large businesses identified a breach or attack, with the average cost to large businesses of all breaches over the period being £20,000 and in some cases reaching millions. The survey also shows businesses holding electronic personal data on customers were much more likely to suffer cyber breaches than those that do not (51 per cent compared to 37 per cent).

The most common breaches or attacks were via phishing emails (tricking staff into revealing passwords or financial information, or opening dangerous attachments), followed by viruses and malicious software (including ransomware).

The survey shows that 'businesses across the UK are being targeted by cyber criminals every day and the scale and size of the threat is growing, which risks damaging profits and customer confidence'.

Of the businesses which identified a breach or attack, almost a quarter had a temporary loss of files, a fifth had software or systems corrupted, one in ten lost access to third-party systems they rely on and one in ten had their website taken down or slowed.

Small businesses can also be hit particularly hard by attacks, with nearly one in five taking a day or more to recover from their most disruptive breach.

Nevertheless it is difficult to get a sense of scale of cybercrime in the UK: citizens often do not know where or how to report a cybercrime; and there is a corporate reluctance to admit cyberattacks (often for reputational reasons). Added to that, the secrecy surrounding nation state to nation state attacks – all results in significant under-reporting of the problem.

However, from the annual surveys from UK Government and Office for National Statistics and the private sector surveys on corporate cyberattacks from the likes of KPMG[26] and PwC,[27] it is clear that the scale, pace and sophistication of cybercrime and cyberattacks are growing and attempts to slow this increase are so far not working.

13.7   Non-cyber Comparators

Before we can search for solutions to a rising cyberthreat landscape, it is perhaps useful to look at a non-cyber comparator – and one with personal relevance to me.

---

[25] https://www.gov.uk/government/statistics/cyber-security-breaches-survey-2017

[26] https://home.kpmg.com/ro/en/home/media/press-releases/2017/06/kpmg-cyber-security-benchmark-2017.html

[27] https://www.pwc.co.uk/issues/cyber-security-data-privacy/insights/global-state-of-information-security-survey.html

I had been seeing friends in Norwich when I returned to the car park at about 10 pm. Keys in hand, I wandered through the well-lit open-air 'Pay-and-Display' to where I had left my car. As I was three or four parking bays away from my own space, the confusion set in, followed by disbelief, and then the slow realisation that my car was not there. It had been stolen. It was Thursday 19 June 1986.

According to the Office for National Statistics (ONS), my stolen Ford Capri was one of 988,000 vehicle and vehicle-related thefts reported to the police in 1986.[28] The Crime Survey for England and Wales estimated that over 2.7 million instances of vehicle-related theft took place that year. Moreover, what subsequent years would show was that this type of crime would continue to rise, peaking in 1993 with 1.5 million vehicle-related thefts reported to the police and a Crime Survey for England and Wales[29] estimate of 4.3 million vehicle crimes having taken place during the year. In 1993, of every 100 vehicle-owning households, 20 were the victims of a vehicle-related theft.

As my experience suggests, breaking into and stealing cars was relatively easy in the late 1980s – virtually anybody who wanted to could do it. Coat hangers, screwdrivers and even tennis balls were used by thieves to unlock doors or override central locking mechanisms and break ignition systems and steering locks. The contents of the car could then be pilfered or the ignition hotwired and the vehicle driven away either for resale, scrappage and spare parts or just joyriding. It was the cybercrime of its day.

But 25 years later in its 2017[30] bulletin on Crime in England and Wales, the ONS reported that around four in 100 vehicle-owning households were the victims of vehicle-related theft. This means that in a generation, we have seen a reduction of 80% in this type of crime. Even in a society with a reducing acquisitive crime rate overall, this is a significant change.

Public education no doubt had a part to play: the TV and poster campaigns of the early 1990s providing crime prevention advice – encouraging us to lock our cars, not leaving valuables in plain sight and putting bags in the boot to reduce the temptation for passing opportunist thieves. But it was also about this time that vehicle manufacturers began to design-in increasingly sophisticated security features: better keys/locks, removable radios and CD players, built-in alarms and immobilisers as standard. Later in the twenty-first century, we see even more sophisticated entry systems and tracking devices.

Arguably the catalyst for this was the decision in 1992 for the UK Home Office to publish its 'Car Theft Index'.[31] For the first time, the best and worst models for security were made public – and unsurprisingly the public made it clear to the

---

[28] https://www.ons.gov.uk/peoplepopulationandcommunity/crimeandjustice/articles/overviewofvehiclerelatedtheft/2017-07-20

[29] http://webarchive.nationalarchives.gov.uk/20110218145224/http://rds.homeoffice.gov.uk/rds/pdfs2/hosb1996.pdf

[30] https://www.ons.gov.uk/peoplepopulationandcommunity/crimeandjustice/bulletins/crimeinenglandandwales/june2017

[31] http://library.college.police.uk/docs/hopolicers/fcpu33.pdf

manufacturers that they wanted a car that was harder to steal by *not* buying those vehicles featuring at the top of the Index. As a result, the risk of car theft reduced.

So the question is, are there ways we can reduce the risk of cybercrime and increase cybersecurity, and if so how?

13.8 Understanding Risk

The UK Government's national risk assessment for 2017[32] highlights that criminal gangs operating in different parts of the world have 'growing capabilities' when it comes to 'holding entities to ransom' after they had succeeded in encrypting business and personal data (i.e. criminals are getting better at using ransomware). As a result, the UK Government now ranks this particular type of cybercrime alongside terrorism and certain natural disasters as an important risk for the UK.

In the physical world, we tend to close windows and lock doors on leaving our homes and cars. On the whole we avoid unlit alleyways in certain parts of town we know or believe to be less safe. And we would generally think it unwise to leave valuables unattended in a pub or a restaurant. Industry protects its physical and intellectual property with CCTV, human guards, alarms, keeping corporate secrets under lock and key, and vetting and screening staff. And nation states have laws and law enforcement and borders and military.

We have a risk-based approach to security: we expect certain enhanced security measures in a bank or jewellers shop as opposed to a greengrocers; and it would be odd indeed to go to a high-profile event or even a large shopping mall and not to see security staff helping to ensure public order and safety. Similarly, we expect officials at our ports and airports checking documents to ensure that those deemed to cause us harm are kept from entering the country.

This is because the idea of risk, whether personal, corporate or nation-related is not innate but is a calculation based on three things: our assessment of the *threat* we are facing (from burglars, car thieves, corporate espionage, fraudsters, terrorists or indeed anything else from climate change to disease) compared with our *vulnerability* to that threat – our susceptibility to whatever it is we are facing. The third aspect of risk is *impact* – how much harm we think might be caused if the event we are considering actually occurred. We can think of this like an equation: risk = impact x likelihood (where likelihood = threat and vulnerability).

The laptop bag in the front seat of our parked car would increase our *vulnerability* to theft in an area where there is a known *threat* from thieves (and therefore heightening the *likelihood* of the car being broken into), but the fact that we had removed the laptop from the bag before leaving the car would reduce the *impact* of the crime.

---

[32] https://www.gov.uk/government/publications/national-risk-assessment-of-money-laundering-and-terrorist-financing-2017

Together, *threat* and *vulnerability* help form our understanding of the *likelihood* of an event happening. But *threat* and *vulnerability* are not constant and vary according to many different factors. As a result *likelihood* also varies, and it is generally accepted that if numerical values are placed upon *threat* and *vulnerability*, the *likelihood* will be always be equivalent to the lower of the two.

As an example, if my Ford Capri is parked unlocked (high *vulnerability* to theft) outside my house on Uist in the Outer Hebrides of Scotland (very low *threat* of theft), then the *likelihood* of my car being stolen is very low. Conversely, a more modern and secure car parked in a locked garage (*vulnerability* low) in a relatively high-car theft area like central London or Manchester (*threat* high) would also have a low *likelihood* of theft.

Understanding this process allows us to step back from the cyber-related maelstrom of almost daily stories of data breaches, hacks and often harrowing accounts of lost identities and stolen assets. In order to increase our cybersecurity, we need to reduce the risk of cyberattack and cybercrime. For this to happen, we have to reduce either the cyberthreat or reduce our vulnerability to cyberattack or both.

13.9 Reducing the Cyberthreat

All of us are potential victims of cybercrime, but with a widespread and global cyberthreat from nation states to hacktivists to organised crime groups, there is little the citizen or the corporate entity can do to reduce directly the threat from the 'actors' involved – threat reduction is almost exclusively the domain of the Government and law enforcement, and the UK has made some movement in the direction of tackling this with a National Cyber Security Centre (NCSC) focused on reducing the threat as well as the vulnerability of corporate victims in the UK.

At the launch of the NCSC in London on 14 February 2017,[33] the Chancellor of the Exchequer, Philip Hammond said:

> 'But as we enter the so-called "Fourth Industrial Revolution", we have to be alive to the fact that this transformation is not without its challenges. The development of artificial intelligence heralds a technological revolution that will fundamentally change our lives. But it will also disrupt existing patterns of work, life, and society.
>
> 'The fact is that the greater connectivity that will enable the development of the digital economy is also a source of vulnerability. And those who want to exploit that vulnerability have not been idle. The cyberattacks we are seeing are increasing in their frequency, their severity, and their sophistication. In the first three months of its existence, the NCSC has already mobilised to respond to attacks on 188 occasions.
>
> 'And high-profile incidents with Sony, TalkTalk, and TV Monde have reminded us of the scale of damage that a single successful cyberattack can inflict. So this new centre, and its work, is vitally important'.

But it is very unlikely given the globalised threat from cyber that any one of our law enforcement agencies will arrest their way out of the cyberthreat we are facing.

---

[33] https://www.ncsc.gov.uk/news/chancellors-speech-national-cyber-security-centre-opening

Cybercrime is a multifaceted and multinational threat with many different attack vectors. Successful UK law enforcement response requires ring-fenced and specialised policing if it is to investigate and successfully prosecute those individuals that are within its jurisdiction – and much of that expertise is currently within the private sector.

New and evolving threats require a similar response from those facing them. A successful law enforcement approach to cybercrime requires not just a behaviour change in ring-fencing resources for cyber-related work, but also training police officers to be more proficient in investigating and collecting evidence on a crime where victim and criminal have not physically met. It also requires a *culture change in law enforcement* in using the facilities and expertise of the private sector to collect, collate and analyse sometimes terabytes of data in order to determine the evidence of a crime.

13.10   Increasing Cybersecurity by Reducing Vulnerability

If we have to rely on the state to tackle the threat, then as citizens, we have to focus on vulnerability reduction.

'You don't have to run faster than the bear to get away. You just have to run faster than the guy next to you'[34] – Jim Butcher's quote is particularly relevant in the world of cybersecurity. If you can make your system or network less vulnerable than somebody else's, the attacker will go along the path of least resistance. With a global cyberthreat, a cybersecure country or city will direct the attacker to focus on other cities, countries or continents.

In addition to tackling the threat, the nation state has an obligation to help advise it's corporate and citizen inhabitants in reducing their vulnerability to cyberattack. Advice and guidance in helping to change online behaviour is a key part of that and will be informed and influenced by the state's understanding of the sophistication of the threat. Having clear and efficient communication channels between those tackling cybercrime and those defending against it (or helping people to defend against it) and *sharing knowledge on data and good practice* is a vital part of vulnerability reduction. This can be via online portals, broadcasting, advertising and citizen education and awareness programmes, outreach from government itself and/or cyber-specialised law enforcement officers working as a network of cybercrime reduction teams. Just as with car crime in the 1990s, citizen education is an important part of vulnerability reduction.

In the cyberworld, because data is often the most valuable asset a person or a company can hold, it is surprising how many of us do not know what data there is about us online or – in the case of corporates – how much data we might hold, where it is stored or what the consequences might be of losing it. According to one piece of research published in March 2016 by Veritas Technologies of California, the

---

[34] http://www.jim-butcher.com/jim

average business has visibility of 48% of its total holding of data (and a further 33% was considered 'redundant, obselete or trivial'). As a business, it would be prudent to have a *data audit to examine how much data is held*, determine what is and what is not necessary for the purposes of running the organisation, securely delete the rest and establish the value of the remainder: what would be the impact if the data or IT systems were to be compromised and ensure that it is protected accordingly.

As a citizen, having a data audit is also a positive step in reducing cyber-vulnerability. Regularly 'Googling' yourself is not vanity, it is good cybersecurity practice. Determining and editing how much information you have shared about yourself and your friends and family on social media will make it harder for organised criminal groups to send spear-phishing emails or indeed to steal your identity and apply for goods, services and credit in your name. But knowing how much of your data is available for sale on the dark web – often as a result of data breaches – is a more difficult, if not impossible, task. From May 2018, companies operating in the UK and EU will have a legal obligation to alert consumers (and indeed the authorities) immediately to a loss of data or breach of corporate cybersecurity and then help their customers to protect themselves: perhaps with credit report monitoring through one of the credit reference agencies or through the use of an online identity protection service.

There are also certain internet-enabled activities which, if not managed securely, can lead to increased vulnerability of cybercrime and cyberattack. The oft quoted 'Internet of Things' is the interconnection of computing devices embedded in everyday objects, enabling them to send and receive data. These devices are now part of our lives from the devices we wear (for medical and leisure purposes) to our homes and the vehicles we drive. They are also found in corporate networks including industrial control systems (widely used in sectors such as energy, water, transport, manufacturing and pharmaceuticals) and building management systems (which are used to control and monitor the mechanical and electrical equipment in most modern buildings such as ventilation, lighting, power, fire and security systems). The Internet of Things allows a device to sense and monitor the environment in which it is operating. For this reason it is critical to try and *protect interconnected devices from interference* by cyberattack by password protection, removing unnecessary functionality and ensuring that the system is regularly updated (or patched) with new software releases from the manufacturer.

Organisations have a significant part to play as has already been mentioned, not only in reducing their own cyber-vulnerability but also helping to protect their customers and clients as well as their employees. The importance of a cybersecurity risk management approach originates from the boardroom and should filter through the organisation with appropriate policies and culture affecting corporate behaviour towards employees, contractors and supply chain. A company should clearly state its cybersecurity approach internally to its staff but also publicly to its customers and consumers.

This means *having a cybersecurity culture*: ensuring systems are updated regularly, achieving cybersecurity accreditations such as ISO27001 and the UK Government's Cybersecurity Essentials, requiring similar accreditation for its suppliers, educating staff about the dangers of phishing attacks and the importance of a cybersecurity culture in the office and at home (because so many of us take work home and/or work from home). reducing privileged user access to your systems and reducing USB port access to those employees who really need it, having 'protect and detect' anti-virus software and knowing what to do if the worst happens and you are subject to a cyberattack and monitoring usage of your systems to alert you to any unusual activity.

But having a cybersecurity culture is also applicable to the consumer citizen too. We should all regard as part of everyday life: being wary of unsolicited emails and telephone calls asking for bank or shopping account details; looking out for phishing or spear-phishing activity and not clicking on links or attachments that look suspicious; not using free public Wi-Fi hotspots for banking or shopping but only browsing – criminals create fake Wi-Fi hotspots in order to capture personal data, identities and your account login information; and making sure your computer or device is up to date with software releases and login passwords are strong and not used for multiple accounts.

Earlier, I said that cybersecurity is as much about the security surrounding people and physical infrastructure as it is about the online and digital environment because *cybersecurity is, after all, security*. Any or all of the above activities helping to reduce corporate vulnerability to cybercrime will be undone by having a lax security approach to vetting or screening staff, employees and contractors as it will be leaving the office unlocked out of hours. Cyber vulnerability reduction has to be seen in the wider context of corporate security because criminals will use all the tools at their disposal to get what they want from you. Tailgating a genuine employee will get a criminal past the most sophisticated swipe card or keypad entry system, and if a stranger in the office is not (politely) challenged, then all manner of cybercrimes can be conducted from the inside through the application of keyboard loggers or the injection of malicious software in to your network via unsecured USB ports. Similarly, although not common, the criminal insider can cause havoc by dismantling or overcoming all manner of security procedures as well as having privileged access to data and employees. For the citizen, the equivalent is having a fully patched and password-protected laptop with anti-virus software left in view from an open window and subsequently stolen.

13.11   Thinking Differently about Cybersecurity

One of the most common cybercrimes affecting the citizen is identity crime, yet we have used the same concept of identity[35] for nearly 150 years: name, address and date of birth. This is looking increasingly antiquated given the pace of our digital lives and the fact that we apply for most goods and services increasingly online and not in person over a counter. Even if we extend the data capture used by many banks

[35] https://www.cifas.org.uk/insight/fraud-risk-focus-blog/changing-nature-identity-whats-in-a-name

for identity authentication and verification to name, gender, address, date of birth, mother's maiden name and telephone number/email address, of the six data fields, four of them (name, gender, address and telephone number/email address) are changeable with varying degrees of effort. We need to look at what constitutes identity in the digital age.

A lot of attention has been given to the use of biometrics: authentication based on a human characteristic unique to the individual which, in theory, cannot be copied. Examples of biometrics include voice patterns, face, fingerprints and veins in the finger. However, biometrics is not without flaws: there are debates about security, and currently only one access channel can be used at a time: voice recognition works over the phone; vein scanners only work online; and fingerprint readers only work when a person is trying to access their account through a mobile banking application on a device.

If not biometrics – or not solely biometrics – we need to look elsewhere: a layered approach using blockchain technology, biometrics, social media scraping (collating all available information about you from your social media accounts) and open-source information. The technology is available, but the catalyst is still not there.

In reducing our corporate and individual vulnerability to cybercrime, we need to change our behaviours, but we also need a cultural shift in how we view the threat of cyberattack. Ask any millennial, and they will view without distinction the online and offline world – it is one world with different dimensions. This in part goes to explain why the under 21 year olds are the fastest growing victim group for identity theft[36] – they share everything and lead their lives online which plays in to the hands of cybercriminals and fraudsters. We are unlikely to be able to change their attitude to the limitless potential of an interconnected world – instead we need to improve security and safety in this digital age by having the cyber equivalent of the Car Theft Index of 1992 – we need a catalyst for cybersecurity.

Even in the 1980s, it would be unheard of when buying a car to be asked whether we wanted brakes, headlights or a reverse gear as an optional extra. But every time we buy a laptop or interconnected device, we have the option to have or not to have a password to access the desktop, anti-virus software, encryption and all manner of other add-ons which would make our online activity safer. Building-in cybersecurity to our devices and computers and then having an independent index of their effectiveness would provide a guide to the consumer – and, as with vehicle security, why would anyone buy a laptop knowing that it was less cybersecure than another one at a similar price?

Finally, we need to look back again at the state. The engineering of the Internet allows many countries to monitor and potentially block unwanted or unwelcome traffic to their citizens: North Korea and China do this particularly effectively albeit with a detriment to free speech and regime opposition. But state-backed shielding of citizens and companies does not have to be totalitarian. With suitable checks and

---

[36] https://www.cifas.org.uk/insight/fraud-risk-focus-blog/educating-young-people-about-fraud-key-protecting-them-online

balances and proper effective democratic oversight, the state too can play a more effective role in protecting its inhabitants by blocking at the front door much of the malicious code, phishing attacks and DDoS originating from overseas - in doing so ensuring our greater safety and security in a digital age.

Chapter 14
A Sustainable Higher Education Sector: The Place for Mature and Part-Time Students?

Suzanne Richardson and Jacqueline Stevenson

14.1 Introduction

In England, explicit policy efforts to improve the access of those groups under-represented in higher education (HE) have formed a key element of national education policies for over 50 years. Successive waves of expansion shifted English HE from an elite system to a mass one with institutions such as the Open University and Birkbeck College and more recently the further education sector, playing a significant role in expanding the access to HE of part-time and mature learners. Over the last decade, however, there has been a 60% drop in part-time student numbers and a 50% drop in the number of mature learners, as a result of, among other factors, the shift from grants to loans, the introduction and then the rapid raising of fees, and sector-wide institutional reprioritisations leading to the slashing of part-time course. At the same time, it is estimated that 80% of new jobs will require degree level qualifications by 2020, a target which cannot be met solely by those 40% of young people who access HE. This posits the question of how such economic goals are to be met without the reinvigoration of access for part-time mature learners. As Hillman (2015, p. 4) notes,

> The collapse in part-time study is arguably the single biggest problem facing higher education at the moment. There are other challenges too, such as the future of the research environment, how to assess the quality of teaching and dealing with the effects of marketisation. But it is the fall in part-time learning that is probably the biggest black spot.

More recently, however, the growth of higher apprenticeships and degree level apprenticeships is progressively bringing part-time learners back into HE. Both groups of students are different from the traditional student who studies

S. Richardson (✉)
Leeds Beckett University, Faculty of Arts, Environment and Technology, Leeds, UK

J. Stevenson
Sheffield Hallam University, Sheffield Institute of Education, Sheffield, UK

© Springer International Publishing AG, part of Springer Nature 2018
M. Dastbaz et al. (eds.), *Smart Futures, Challenges of Urbanisation, and Social Sustainability*, https://doi.org/10.1007/978-3-319-74549-7_14

full-time and is generally under 21 years of age, and retention and belonging present major challenges for institutions who have focussed primarily on full-time undergraduates.

In the first part of this chapter, therefore, we will chart the rise and fall of mature and part-time student provision. In the second part, we will draw on our own research to present the challenges facing institutions in helping these students develop a sense of belonging, fundamental to student retention and success. We will end this chapter by positing a set of practical recommendations designed to support institutions working to support mature and part-time students.

14.2 The Rise and Fall of Mature, Part-Time Student Numbers in English Higher Education

Under successive Labour and Conservative governments, English higher education in the latter part of the twentieth century was characterised by a period of enormous expansion – the 'massification' of higher education (Langa Rosado and David 2006). Pressure to expand higher education followed the increased post-World War II birth rate, as well as an expansion in those achieving the university entry standard (Ross 2001a). The Robbins review of higher education (Robbins 1963) led to the building of a range of new universities in the 1960s resulting in a huge expansion in the number of full-time university students (Ross 2001a), whilst the creation of 30 new polytechnics in the 1970s – aimed at accommodating students 'unable' to access traditional universities – led to a massive expansion of higher education, with particular increases in local, ethnic minority, female and crucially mature and part-time students (Ross 2001b).

By the late 1990s, the UK university population had shifted from an elite system (defined by Trow 1974, 2005, as enrolling up to 15% of the age group) to a mass one (of between 16% and 50%). Under the New Labour government which came into office in 1997, a commitment was made to achieving universal (50%) participation among 18- to 30-year-olds by 2010 (HEFCE 2006). Although much of the thrust of this expansion saw growth in the full-time sector, it led to a further wave of growth in part-time throughout the end of the twentieth century.

Although the numbers of students accessing full-time higher education has continued to grow, however, there has been a radical and dramatic fall in the number of part-time students. Analysis of information published by the Higher Education Statistics Agency (HESA) by London Economics (Conlon and Halterbeck 2015) shows that between 2001/2002 and 2013/2014 whilst there was a 23% increase in the number of full-time undergraduates, the number of part-time undergraduates declined by 42%. The numbers have continued to fall with the latest figures from HESA showing that part-time student numbers in England fell from 243,355 in 2010–2011 to just 107,325 in 2015–2016 (HESA 2017).

Much of the blame for escalation in the decline can be attributed to financial concerns and the lack of financial support for part-time students. It is perhaps unsurprising that part-time enrolments in England were at a low point in 2012 when the government raised the cap on part-time fees, doubling and in some cases even tripling the cost of many courses. In 2011–2012, before tuition fees went up, just under half of all new students were full-time undergraduates. In 2011/2012, 31% of all enrolments were part-time. By 2014/2015, this had reduced to 25% of all enrolments, and in 2015/2016 part-time enrolments accounted for just 24% of all enrolments (HESA 2017).

In part, the dramatic post-fee hike fall in part-time student numbers is, as Callender evidences (2017), based on the government's flawed belief that employers pay for part-time students who are in work. The reality is that employer sponsorship for part-time degrees has declined significantly (by 35% since the 2012/2013 reforms). Instead the vast majority of part-time students are mature, adults who are already in the workforce, who are combining higher study with a job, and have ongoing obligations to family such as mortgages and other significant financial commitments. This makes them unwilling or unable to take on large additional amounts of debt. Moreover, up to two-thirds of those who might study part-time are not eligible for student loans (Callender and Mason 2017). There were many exemptions to the loans. Those who already hold a degree are not generally entitled to them nor are people studying a module or two, rather than a full qualification. In total, only a third of part-time students in England are entitled to tuition fee loans, and even fewer have taken them out. The other two-thirds face upfront fees. Applications to full-time undergraduate courses for over-25 s fell particularly steeply – by 18% in 2015 – confirming a general trend of dwindling participation in HE among adults.

As Callender (2017, p. 18) notes,

> The falls have been greatest among older students, especially those aged 55 and over, those with low-level entry qualifications or none at all, and those studying less than 25 per cent of a full-time course who choose bite-size courses.

In summary, the Independent Commission on Fees said raising the cost of undergraduate tuition to £9000 a year has led to 'a significant and sustained fall in part-time students and mature students'. It added: 'We believe that the new fee regime is a major contributory factor' (The Sutton Trust 2014).

These changes to part-time, and mature, student numbers are not only highly relevant to the UK economy but represent a significant social change. Part-time study helps upskill and reskill the workforce, supports economic growth, promotes social mobility, helps to build society, and allows disadvantaged individuals to improve their potential. The collapse in part-time and mature students studying at universities in England therefore threatens social mobility and economic performance and needs to be addressed urgently if there is to be a boost in productivity and a rebalancing of the economy. Without providing the opportunity to retrain or improve existing skills, both will be difficult. Overall, the higher education system is becoming less diverse, less accessible to older adults, and less relevant to the challenges of modern society (Richardson 2017). All of this, it should be added, has been an entirely predictable result of successive government policies. It has been well known for some time that part-time higher education does not appear to be a priority for any recent government.

As most part-time mature students tend also to come from less well-off, nontraditional backgrounds, this decline has also had an impact on the social mix of our universities and on efforts to widen participation. Typically widening participation students are those who want to do 'bite-size' courses or part-time study, and the fall in numbers has been greatest among these older students with low level entry qualifications. Unsurprisingly, it those with the greatest commitments and other responsibilities who have found themselves excluded by the increase in fees and the introduction of loans. Even if the loan entitlement were not so tightly constrained, it would not solve all the issues, mainly because part-time students are more price sensitive than full-time ones but also have many more responsibilities than the average school leaver at 18 years old.

The current fall in demand means universities are quietly closing their part-time courses. Institutions specialising in part-time study have seen dramatic declines. Birkbeck, part of the University of London, which specialises in part-time study, has been forced to re-engineer its courses to offer intensive evening courses which enable students to gain a degree in 3 years. The Open University, which, historically, solely delivered distance learning and relies on older students who are already in employment, has been hit especially hard, with its numbers falling by 30% between 2010–2011 and 2015–2016 (Conlon and Halterbeck 2015).

As demand declines, the supply, of course, dries up. Universities for whom part-time provision is marginal business are not surprisingly closing courses. Indeed as Callender (2017 p1) has said,

> The Russell Group has pretty much pulled out of part-time undergraduate education,' and that 'these course closures matter more for part-time students than full-time ones, because most part-timers are juggling a job or a family, or both, and so need to study near home… Lack of local provision won't mean they will try studying somewhere else; they just won't study at all.

Callender (2017) warns that once universities have stopped offering part-time courses, the tap can't simply be switched back on: 'Once universities lose the infrastructure it's hard to reinstate it'.

With UK employment now at its highest level since 1975, almost 75% of 16–64-year-olds are now in work (ONS 2017). At the same time, however, there are more unfulfilled job vacancies than ever previously recorded (777,000) and a severe skills shortage, especially in STEM occupations (ONS 2017). Research by EngineeringUK (2017) suggests that an additional 1.6 million engineers and technically qualified people are needed by 2025 as currently demand outstrips supply. These issues may only be exacerbated by the potential effects of the UK's prospective withdrawal from the European Union (Brexit) as there is a high dependency on attracting and retaining international people. The current Conservative government has, reassuringly, begun to show signs that it recognises this skills crisis with the publishing of the Industrial Strategy Green Paper (Dept for BEIS 2017), the introduction of the apprenticeship levy and a renewed look at technical education through institutes of technology that are expected to increase the provision of available higher-level technical education. The success of these interventions remains to be seen. These initiatives are perhaps unlikely to fully address the collapse of part-time study in higher education which is considered an essential component in addressing the skills shortages.

It is readily accepted that part-time study provides an invaluable route into higher education, for individuals who may not have had the opportunity to attend university straight out of school and now require the flexibility to continue their education part-time whilst meeting work and family commitments (Burke 2012). Despite this, the fall in mature, part-time students raises questions for the current debate about how the sector will provide lifelong learning opportunities. Few would argue that an element of the fall in demand may be partly due to workers not needing to learn new skills during their careers. On the contrary, many believe that technological change is increasing the rate at which skills need to be upgraded. This is recognised in the Industrial Strategy Green Paper (Dept for BEIS 2017), but it is not certain what role higher education providers will play in meeting this demand or what their offer will be to older workers who want to study at low intensity. The new industrial strategy provides an opportunity to drive growth and raise productivity and at the same time support people who need or want to upskill throughout their working lives. The strategy indicates the government is open to 'exploring ambitious new approaches to encourage lifelong learning, which could include assessing changes to the costs people face to make studying for these qualifications a little less daunting' (Dept for BEIS, 2017 p 45). The green paper makes much of the role of adult skills in the post-Brexit economic renewal and demonstrates an awareness of the need to ensure better articulation between the demand for skills and their supply. Higher-level and degree apprenticeships may become attractive options for mature students, but it could be that many providers mostly offer courses for the already highly skilled. It is also going to have increasing importance in achieving the need to undergo retraining as the rate of change in many sectors continues to escalate.

14.3 New Approaches to Student Support

The decimation of part-time provision has resulted in a higher education sector which is now focussed on attracting, retaining, and supporting young full-time students. If patterns of participation are to change and part-time student numbers are to recover, HE institutions will need to rethink how they enable equitable success for part-time students.

There is manifest evidence to suggest that mature or part-time learners have difficulty settling into university life during their first year as a result of a 'lack of connectedness and involvement', 'loneliness', or 'unhappiness and dissatisfaction' (Perry and Allard 2003). Mature students have complex needs when they enter higher education including low levels of confidence and belief in their own ability to learn but also have the potential to benefit enormously from study later in life through enhanced career opportunities and personal fulfilment. Mature, part-time students tend to have different characteristics and needs to younger, full-time students, particularly in relation with entry qualifications, social background, lifestyle, and time available for study (Million+/NUS 2012). Therefore, students from nontraditional backgrounds may be disadvantaged by not fully engaging in university and/or college social and academic environments. This might be as a result of mature, part-time students being grouped as 'others' who are either deterred from entering higher education for the reasons discussed previously, namely, financial, or who go through university never really feeling as though they belong. Such students are also regarded as being 'unprepared' for learning or lacking academic ability and, so for these reasons, more likely to withdraw before they complete their course (Thomas 2002). It is unsurprisingly, therefore, that many mature individuals consider higher education to be 'not for people like them' and can often perceive themselves as 'not belonging' as a result of feeling they lack either capacity or previous educational experience (Burke 2004).

Part-time, mature students generally fit study around other commitments and are time poor, as opposed to full-time students fitting work around study commitments (Universities, UK 2013). Together, these factors make it more difficult for students to fully participate and integrate, to feel as if they belong in higher education, which can impact on their success. Therefore, institutions need to recognise this and support practices that develop a sense of belonging with part-time, mature students as this has a direct impact on retention and success.

Developing a sense of belonging can however present challenges when institutions develop models of delivery that cater for a more heterogeneous student population (Rowley 2005). A key feature of nonengagement with institutional approaches is the feeling of not 'fitting in' or having no sense of belonging. In other words, the relationship between the student, their peers (social inclusion), and their learning environment (academic integration) has direct influences on retention and success and on the broader sense of belonging which requires institutional commitment. These factors are particularly relevant where belonging is critical to success.

14.4   Challenges Facing Institutions in Helping These Students Develop a Sense of Belonging

A sense of belonging can have many meanings for different individuals, at different times, and in a variety of contexts. Most meanings include the following: belongingness, acceptance, support, relatedness, affinity, and membership (Burke 2012; May 2011). They all deal with the level of integration a student feels to a particular context. This integration includes social involvement, academic connections, and institutional support and commitment which can positively influence retention and success. As Hurtado and Carter (1997) imply, a sense of belonging is more meaningful for those who 'perceive themselves as marginal to the mainstream of [university] life'. It is clear that mature, part-time students feel belonging in a variety of ways and at fluctuating levels, with any perceptions of belonging continually changing and highly dependent on the situation students find themselves in. Students' academic and social involvement can often influence their sense of belonging through regular interactions and frequent connections with others during their studies. Yet for mature, part-time students with diverse previous experiences, multiple identities, work experience, and carer responsibilities, all this can position them on the sidelines of the academic experience, which then limits access to any means of belonging. Part-time, mature students generally fit study around other commitments and are time poor, as opposed to full-time students fitting work around their study commitments. These are factors which make it more difficult for mature, part-time students to fully participate and integrate, to feel as if they belong in higher education, which can impact on their success. Therefore, institutions have to recognise this and provide supportive practices that develop a sense of belonging for part-time, mature students.

Belonging also presents other challenges when institutions develop models of delivery that cater for a more heterogeneous student population. A key feature of nonengagement with institutional approaches is the feeling of not 'fitting in' or having no sense of belonging.

Having institutional supportive practices can help all students, but particularly mature, part-time students, to balance the demands of work with study and personal commitments. Yet in trying to manage these, it can still exclude them from important networking opportunities that offer social support and integration. Multiple commitments and responsibilities outside of higher education study challenge the necessity for social involvement and academic integration within retention and success literature (Thomas 2012). This friction is often a key source of stress for mature, part-time students, even when they set aside academic priorities in order to meet more urgent demands. Although a major benefit to studying later in life is that the students are proactive, highly motivated, and committed to their studies, any preconceptions they may have will either facilitate or impede their learning experience in higher education. Despite significant widening participation activity, part-time, mature students from nontraditional and disadvantaged groups continue to face considerable challenges and can, at times, appear invisible. This obviously impacts on their learning experience and ultimately their progression which is in part due to their diverse backgrounds but also limited experience of higher education. By implication, this can then be linked to social interaction and integrated academic experiences which help connect students to the higher education learning environment. Thus, habitus influences the way in which students integrate and engage with both the institution and the subject and can contribute to their overall satisfaction, progression, and retention.

As Braxton and Hirschy (2004), draw on Tinto's theory of integration (1993), they suggest, students must be socially and academically connected to the institution, and that there is a need for students to be engaged with the institution and committed and involved, if they are to persist in their studies. Positive behaviours can lead to connectedness with others, and this creates feelings of being valued and respected. These academic and social integrations help to connect students to the wider institutional environment and provide a sense of belonging which contributes to their overall success. In contrast, lack of social integration can result in underperformance or withdrawal, mainly as a result of having to cope in an unfamiliar culture, which can be confusing for mature, part-time students who then feel estranged or just 'out of place'. However, despite part-time, mature students remaining a significant section of the higher education student population, the dominant narrative of belonging remains modelled on full-time, younger students. This is a significant omission, since research by Read et al. (2010) has evidenced that mature, part-time students feel less of a sense of belonging than traditional, younger students. Moreover, belonging is more than just an individual feeling; it also has collective consequences in relation to student satisfaction, the achievement of academic goals, student retention, and academic success. Some students can feel lost within a higher education culture for a number of reasons. These might be as a result of the learning support infrastructure, practices accepted within higher education, and/or the delivery of learning and how this differs from previous learning experiences. Mature students often express strong self-doubt about their worthiness at studying in higher education (Reay et al. 2010). It is also the case that the majority of widening participation students are located at less prestigious universities and further education colleges (Universities UK 2015b). Institutions delivering higher education are expected to provide positive experiences for nontraditional students with student expectations and outside commitments likely to influence their engagement. This has been identified as 'a problem for institutions if they only provide traditional teaching opportunities, as students may not attend' (Thomas 2012). When classroom delivery considers work and life experiences and draws upon these in assessment, learning, and teaching practice, then mature students are more likely to feel involved and to have feelings of perceived belonging.

This cohort of students bring with them a wide variety of relevant experiences of the vocational subject area and can add depth to class discussions and clarity of application of theory to work-related learning situations (NUS/Million+ 2012). This can prove to be extremely valuable in the classroom. The higher education system does acknowledge the various needs of students who require different kinds of experiences and support, through differential strategies such as the development of curriculum models that accommodate a more heterogeneous student body and the provision of tailored academic support. However, these have not necessarily been very effective in helping a wide spectrum of students to succeed in higher education (Warren 2002). In order for higher education to be inclusive, its pedagogy, curriculum structure, content, and assessment must be appropriate for different groups, including for those in part-time, mature student groups. The strategy has to be clear that mature, part-time students require a different experience to that of younger learners, including the structure and delivery of a course but also availability and location (Davies et al. 2002).

In trying to balance the differing demands on their time, mature people see themselves as 'day students'. This then limits their participation in wider university life. It excludes them from important networking opportunities and support groups which reduces their ability to engage in extracurricular activities or to be accepted as a traditional student (Lyons 2011). People generally have a desire to integrate socially, and they want to feel involved and to belong to groups with others who share interests. Conversely, any disruption to normal routines and environments might cause a feeling of non-belonging. Failing to achieve an adequate sense of belonging can have important negative consequences which then impacts on the success of the student. However, not belonging does not always result in undesirable outcomes; it is possible that some students may choose not to belong as some may feel as though they have no need to belong.

Studies consistently suggest that students who experience a sense of belonging in educational environments are more motivated, more engaged with others, and more likely to remain on the course and are able to demonstrate greater self-awareness and confidence (Osterman 2000). Therefore, development of an educational experience, which contains influencing factors that address individual needs and cultural desires to become part of an established group, is key to both social and academic integration and student success. Yet, the institutional practices expected to nurture and support a student's sense of belonging are complex and intertwined with engagement, involvement, and factors which promote conditions which foster belonging in higher education, i.e. institutional fit. It is clear that there are key components of 'belonging' which are closely related to those factors which inform retention and success. These are strongly influenced by the notion of congruency (fit) with the institution and the impact of the institutional environment. Belonging is an accepted term that relates to student engagement and social integration. Thomas argues that belonging in higher education is 'closely aligned with the concepts of academic and social engagement' (2012: 12). It is also defined as 'those experiences that help to connect students to the higher education environment and contribute to their overall satisfaction' (Harvey et al. 2006: 23).

In other words, the relationship between the student, their peers (social inclusion), and their learning environment (academic integration) has direct influences on retention and success and on the broader sense of belonging which requires institutional commitment. Belonging is an interrelated aspect of students' higher education academic experience which includes the development of academic networks and an identification with others through the progression of a course, cohort identity, the department, and academic staff across the university or institutional habitus. Establishing and maintaining positive social relationships bring benefits such as creating feelings of inclusion and identity. Having these connections with others can generate feelings and beliefs of satisfaction and happiness (Strayhorn 2013). Due to the complex and abstract nature of the concept, for mature, part-time students, belonging can be found in different circumstances and between the many people with whom they interact. These interactions inform the extent to which any student will feel as if they belong. They support the perception of integration into the higher education community and the development of a more positive experience.

A considerable challenge for institutions is to ensure students do participate and get involved and that there is an inclusive learning environment which enhances academic engagement. To encourage feelings of inclusion and satisfaction, leading to increased levels of confidence and motivation, there have to be strong connections between peer support, faculty support, and class encouragement and enjoyment (Strayhorn 2012). Thomas (2002) demonstrates that students' new social networks at university often provide support to overcome difficulties which result in decreased feelings of isolation and 'otherness' and encourage the development of a sense of belonging. This includes the need for institutional strategies which monitor and evaluate the student experience, bringing both the academics and the students together. This then enables them to share the learning experience.

Perceived belonging in academic environments has a powerful effect on students' emotional, motivational, and academic functioning from anxiety, distress, engagement, and doubts about self-confidence and their own ability. College outcomes such as retention and degree attainment are strongly related to the efforts that students put into their participation in academic, social learning, and developmental activities. However, as previously discussed, the involvement of mature, part-time students in these wider activities is less likely because of commitments outside of study, family responsibilities, financial obligations, and job liabilities. This creates limitations for integration and connectedness with others which can reduce levels of belonging to their institution and a perception of having less sense of community within the institution. For many mature, part-time students, higher education is viewed as a means to maximise their career potential, yet they have to address these previously mentioned issues, if they are to achieve success. Evans (2002) highlights that students have to be instrumental in forming new relationships in order to shape their experiences whilst at university. When people establish friendship networks, it gives them the feeling of being in a meaningful relationship, which helps them negotiate any new learning situation and overcome feelings of marginality. Any significant changes in their surroundings will likely have an impact on their sense of belonging. However, it is clear that belonging may not be experienced by all students, and nor is it a steady or stable perception. At the start of a degree, students might be minor participants, but with experience, (may) begin to contribute more fully, and 'feel a sense of connectedness in both the social and academic aspect' (Thomas 2002: 435) of the university, which is a significant factor in retention and success. This is the challenge facing institutions in helping mature, part-time students develop a sense of belonging. However, there is no doubt that some of the outcomes of studying higher education for mature and part-time students are that these outcomes do have a profound and continuing effect on social mobility, job prospects, and therefore personal lives for individuals.

Studies consistently suggest that students who experience a sense of belonging in educational environments are more motivated, more engaged with others, and more likely to remain on the course. This stems from being able to demonstrate greater self-awareness and with higher levels of confidence. Therefore, development of an educational experience, which contains influencing factors that address individual needs and cultural desires to become part of an established group, is key to both social and academic integration and student success (Nuñez 2009). Yet, the institutional practices expected to nurture and support a student's sense of belonging are complex and intertwined with engagement, involvement, and factors which promote conditions which foster belonging in higher education, i.e. institutional fit.

14.5   Practical Recommendations Designed to Support Institutions Working to Support Mature and Part-Time Students

Belonging to a community of learners has been demonstrated to be fundamental to the success of mature, part-time students. However, many part-time students do not want to get involved or maybe cannot get involved in the same social activities that the traditional, younger student is attracted to. It is unsurprising therefore that they frequently report feeling isolated from other students, of not belonging, and that the institution does little to create a supportive environment for peer group interaction and learning that suits them.

As our earlier research (Richardson 2017) evidences, the key to encouraging mature, part-time students to remain on course and be successful is for them to have institutional support to build their self-confidence; they already have the necessary determination. When these operate together, they encourage students to feel 'at home' in a higher education environment. This allows them to build effective learning strategies and negotiate a successful path with positive outcomes. Across this group of students, there are a variety of social and personal motivations and influences over the decision-making to begin higher education study. For some, this might be the chance to enhance their career prospects, perhaps a wish to change the direction of their life, and for others the opportunity had only just presented itself. Generally, a change in family or lifestyle circumstances can be the trigger to participation, and this then raises their aspirations; for some this might be a significant life event, such as having children or acquiring a new partner.

Mature, part-time students understand [and hope] that going to university will improve their employability and career prospects, which will bring benefits not just for themselves but also for their families too. Other factors in their lives may include dependents, employment status, family support, and prior educational experience. For these reasons, most students are very aware of the need to remain in their current employment for financial reasons and do not wish to place additional burdens on the family budget. But they are keen to seek ways to alleviate the difficulties they have in balancing work, family, and study commitments, often choosing to study at an institution closer to home and preferring to study in the evenings or at weekends. This establishes the need to explore ways to enhance the learning and teaching environment to improve academic and social engagement and the institutional support available for this group of students. However, all students have different motivations and issues to contend with when they choose to study part-time. All of which are significant considerations for the institution, particularly when enrolments are reducing and greater numbers of mature, part-time students are withdrawing from their courses early and not completing their awards.

The support required by part-time students to help them engage and flourish within higher education is fundamental to their success. This is particularly evident when these individuals currently sit on the periphery of higher education, are already different from the traditional student and when retention and belonging are major challenges for the institution with a focus primarily on full-time students. However, these students have very different perceptions and expectations of the support they receive when they struggle to reconcile competing priorities and feelings of being treated differently from the traditional school leaver, younger student, studying full-time. They tend to be able to negotiate their own sense of belonging and are highly strategic in managing their way through the higher education learning environment. This is fundamental to the ongoing debate about the social and academic factors which help shape mature, part-time students' sense of belonging when studying higher education.

The relationship between having little sense of belonging and institutional commitment is evident from research recently undertaken with a group of mature, part-time learners. Institutional commitment is demonstrated through the actions of the institution and therefore plays an indirect role in influencing the development of a sense of belonging. From the research group, there were a number of concerns raised in relation to the facilities provided by the institution as not being suitable for their needs, in part because they lived off campus and travelled in only for timetabled classes but also through a lack of suitable space for group work, spending free time with peers between classes and on campus cafes not being open in the evenings. Thus, a broader recognition of the need for familiarity and suitability of the spaces students occupy and the structure and the delivery of a course for diverse student cohorts are essential. Specific examples might be a comfortable common room space that is dedicated to higher education students with time allocated in the timetable for group work in a relaxed, suitable meeting place for mature, part-time students. Facilitating this interaction may make the complex learning environment feel more academically supportive which in turn can encourage belonging.

The overall quality of staff-student relationships is highlighted together with the accessibility of individual academic support. The inclusivity of the curriculum and particularly when it is tailored to the needs of these nontraditional students should demonstrate supportive interrelationships and more importantly a student-centred approach to teaching. Rather than an institutional environment which facilitates encouragement and collaborative group discussions, there were instances where the significant past experiences of students were often ignored as an important resource in the classroom or where tutors responded in a patronising tone. The students wanted their prior experiences to be acknowledged and valued, and they wanted to feel as if someone cared about them; this would help them to feel connected to others. Without this, the students felt frustrated and ignored which then created tensions between the operational practice and student expectations and ultimately their own feelings of not 'fitting in'. This previous experience is something which is immensely valuable for the whole group, and tutors need to draw on this expertise. Academics have a key role to play in supporting students and ensuring they are not invisible in the classroom. This inclusion may better help them negotiate their new learning situation and reduce any feelings of marginalisation. For mature, part-time students, the lack of familiarity with academic culture can often result in feelings of confusion which may be compounded by unhelpful comments and feedback in the classroom. This group of students require dedicated structures and academic support networks to ensure they are not forgotten about within the institution. It requires a broad-based commitment from many people across the institution to work together to shape expectations and influence the students' sense of belonging.

Mature, part-time students require both individualised tutor support and an academic support network to enable them to work through academic requirements. Activities which bring students together for discussion and support them with managing multiple roles whilst studying higher education are considered useful in developing a sense of belonging. These elements need to work together; otherwise, for the mature, part-time student, their studies become unsustainable. For most students, having peer support from other mature students, who were in the same situation, can help them through difficult periods. Mature students often sought and valued contact with other mature students. This might be as a result of study requirements or the difficulties they had in combining work responsibilities with study. However, students need to be allowed to decide how and when they interact with others. For some, this contact may be required frequently, but for others, this is less so. There can be students who feel they had no need to belong but then recognise that at times they need and want individualised academic support, usually when they felt under pressure at assignment hand in time or when something went wrong. Mature students appreciate sympathetic tutors and flexibility over deadlines when they have particular situations to deal with at home.

Particular problems arise when assumptions are made that all students are the same and part of the younger cohort, with the same needs and wants as full-time students. Mature, part-time students want to engage infrequently for a variety of reasons and that this should be their choice. Overall, the students felt that the tutors could do more in the first few weeks to reflect on students' prior knowledge and experience. Where this was evident, it helped foster rapport and provide reassurance. Through the development of class-level belonging, a crucial part of the higher education experience, this support improved their engagement with learning but also their confidence and motivation. One practical suggestion is to have tutors who have previously been a mature student allocated to mature, part-time student classes to encourage students as cocreators of learning. Students should be considered as part of the community of learners with flexible teaching that facilitates outside responsibilities without causing undue pressure or constraints on them.

Academics have a role to play in helping students' foster class-level belonging, and this is considered a crucial part of the higher education experience. This does present a challenge for staff as it can be difficult to promote collaboration and social interaction when the students have such a busy schedule of teaching delivered in very condensed periods. The development of an inspiring and understanding relationship with supportive staff and a perception of connection within the class group are two factors which contribute to a sense of belonging. Academic staff including those who act as seminar leaders or course leaders should be encouraged to understand the background experience and motivation for study for each mature student. They can then draw on this information in class but also use this to develop a greater understanding of the needs and expectations of each mature student. With the focus on mature, part-time students having been out of education for some time, this individual support helps to reduce levels of anxiety and frustration and improves the motivation and belonging of the students. Higher education study can be a difficult but fulfilling experience, but when the perceived support or effective processes fall short of expectations, then this can hinder their progress and success.

The exclusion from networking opportunities reduces any ability to feel involved and to belong. Whilst perceived commitment from the institution is critical for students to feel socially included, this still requires the support of key players in the learning environment to facilitate this, including academics and peers. This is not helped by the limited time students spend on campus which restricts their opportunities for social engagement outside of class time. Where there was little sense of belonging to the institution or the course, the students can be creative in their use of social media and informal study groups to encourage belonging. They clearly value peer support from others who share similar difficulties and who were also 'juggling' work and personal commitments with higher education study.

In trying to adjust to university life, mature students know from an early stage that peer support is invaluable in the development of a sense of belonging. Peer interaction clearly increases feelings of being valued and confirms its importance and impact on student success. Despite feelings of exclusion, these mature students successfully manage to negotiate their way through their studies with tenacity and a strong determination to succeed. This highlights the importance of the institution to provide a range of strategies which cater for the nontraditional student which makes them feel included and not forgotten about. All mature, part-time students have particular constraints on their time which require support in a sympathetic and reassuring setting. Where this is not readily available, the students feel uncertain and being out of place. Any absence of belonging can lead to decreased interest or motivation and reduced levels of engagement in academic activities. Failing to recognise these needs has considerable significance and thus implications for institutional policy and practice. A strategy is required to improve the development of peer networks with a particular focus on the opportunities and potential this may contribute to student success and the overall learning experience.

Mature students value social contact with other mature students and would like the universities to facilitate this. If this is facilitated by the institution, it would alleviate some of the pressures students' experience which are mainly related to concerns around academic practice, no one to share anxieties with and in the creation of a positive learning experience. This absence of social contact creates feelings of isolation and bewilderment. Moreover, for many mature students, the experience of higher education is a struggle that stems from a lack of confidence in their own ability and/or the perceived lack of support from academic staff. In addition, students also struggle with the need to reconcile competing priorities and the pressure of juggling demands on their time. From this, it is evident that there are varying notions of belonging with some suggesting it means feeling safe at work or home but with a strong absence of any emotional attachment to the learning environment or the course. Although feelings of exclusion and low levels of belonging are obvious, the students show determination in not allowing these aspects to disadvantage them or to reduce the commitment to their studies. Students regularly demonstrate strong desires to 'fit in' and to get involved. They can be effective in establishing new relationships which helps them negotiate the learning environment, including agreeing access to staff at a time to suit them. This enabled them to deal with the situation of being on the margins of higher education and to negotiate their own sense of belonging using determination and a drive for success. This then enables them to deal with issues of frustration.

The two key factors are academic and social integration, and where processes support these, the likelihood of early withdrawal can be reduced. Institutional practice designed to support all students has a significant effect; however, practice more suited to the traditional, younger, full-time undergraduate student who lives on campus can work against mature, part-time students' sense of belonging. This results in the following fundamental elements:

Relationships with academic staff were an important part of mature students' integration into higher education. Academic tutors who were willing to nurture learning and demonstrated high levels of interpersonal skills contributed to students feeling engaged and motivated which in turn helped the students to feel valued and less frustrated. The university culture has a major role in helping mature students foster belonging through the positive interaction between peers and academics. Where this is missing or individuals perceive the social and cultural practice (such as spaces being occupied or designed for younger, more traditional students) to be inappropriate, then this can result in feelings of not 'fitting in', lowers morale, and impacts on commitment. A consequence of not understanding academic practice, a fear of lacking academic ability or of perhaps feeling unable to adapt to the higher education environment, can influence levels of self-confidence and academic integration. This highlights how mature, part-time students are likely to feel less sense of belonging when they struggle to integrate academically.

14.6 Final Reflections

The provision of part-time education is beneficial for social mobility, encouraging under-represented groups into higher education and allowing for the continual upskilling of the country's workforce. If the Conservative government is serious in this aim and wants to, as suggested in the Industrial Strategy Green Paper (2017), open this opportunity to enhance social mobility and career advancement, then it needs to reinvestigate how the burden of debt could be eased for those seeking to undertake part-time study. Furthermore, there needs to be consideration and a solution which provide support for those caught between their current financial responsibilities and the desire for personal and career advancement. If the government truly wants a highly skilled 'global trading nation', then it will need to support a flexible and diverse higher education sector that is able to deliver it. This could include 'assessing changes to the costs people face to make them less disheartening, improving access for people where industries are changing, and providing better information'. An interconnected industrial strategy, with an 'ambitious new approach to encouraging lifelong learning' (Dept for BEIS 2017 p39) at its heart, is a big step in the right direction. But it will require a major shift in culture to deliver it, with ministers and civil servants looking beyond schools and elite universities. It has to be fully recognised that education is for adults too and needs a long-term commitment to support it. In February 2017, the university minister, Jo Johnson, announced the introduction of 2-year fast-track degrees as part of the higher education bill. He said then that students, especially mature ones, were 'crying out for more flexible courses, modes of study which they can fit around work and life' (Johnson 2017). If institutions are to provide the level of support that these part-time, mature students require, then there has to be a better understanding of their needs, aligned to flexible delivery approaches and differential strategies for those who have other priorities and responsibilities. This is particularly true when most are looking for a mode of study that 'fits' with the demands of balancing work and personal commitments.

Regardless of how and why part-time student provision is reinvigorated in England, however, policymakers, academics, and support staff must also recognise that retaining these students and enabling their equitable success require different pedagogic approaches and retention strategies to those currently being used to support young full-time students.

References

Braxton, J. M., & Hirschy, A. S. (2004). Reconceptualising antecedents of social integration in student departure. In M. Yorke & B. Longden (Eds.), *(2004) Retention and student success in higher education* (pp. 89–102). Buckingham: Open University Press.
Burke, P. J. (2012). *The right to higher education*. London: Routledge.
Burke, P. J. (2004). Women accessing education: Subjectivity, policy and participation. *Journal of Access, Policy and Practice, 1*(2), 100–118.
Callender, C. (2017) *Part time student numbers collapse by 56% in five years*. Available online at https://www.theguardian.com/education/2017/may/02/part-time-student-numbers-collapse-universities.
Callender, C. and Mason, G. (2017) Does student loan debt deter Higher Education participation? New evidence from England. *LLAKES Research Paper 58* Available online at http://www.llakes.ac.uk/sites/default/files/58.%20Callender%20and%20Mason.pdf.
Conlon, G. and Halterbeck, M. (2015),Understanding the part-time RAB charge. In N. Hillman, (Ed.) (2015) *It's the finance, stupid! The decline of part-time higher education and what to do about it*. Available online at http://www.hepi.ac.uk/wp-content/uploads/2015/10/part-time_web.pdf.

Davies, P., Osbourne, M., & Williams, J. (2002). *For me or not for me? That is the question: A study of mature students' decision making and higher education* (pp. 1–95). London: DfES.

Department for Business, Energy and Industrial Strategy, (2017) Building our Industrial Strategy. Available online at https://beisgovuk.citizenspace.com/strategy/industrial-strategy/supporting_documents/buildingourindustrialstrategygreenpaper.pdf

Engineering UK (2017) The 2017 Engineering UK: the state of engineering report. Available online at https://www.engineeringuk.com/news-media/2017-engineering-uk-the-state-of-engineering-published/

Evans, K. (2002). Taking control of their lives: Agency in young adult transitions in England and the new Germany. *Journal of Youth Studies, 5*(3), 245–269.

Harvey, L., Drew, S., and Smith, M. (2006) The First Year Experience: a review of literature for the Higher Education Academy. HEA Haskins, T., Brown, K (2002) Entry routes to success: The relationship between the entry route and success of mature students in Higher Education: Liverpool Hope University College.

Higher Education Funding Council for England (2006) Annual report and accounts 2005-06. London: TheStationery Office.

Higher Education Statistics Agency (2017) Higher education student enrolments and qualifications obtained at higher education providers in the United Kingdom 2015/16. Available online at https://www.hesa.ac.uk/news/12-01-2017/sfr242-student-enrolments-and-qualifications

Hillman, N. (2015) Introduction, In N. Hillman, (Ed.) (2015) *It's the finance, stupid! The decline of part-time higher education and what to do about it.* Available online at http://www.hepi.ac.uk/wp-content/uploads/2015/10/part-time_web.pdf

Hurtado, S., and Carter, D.F. (1997) Effects of college transition and perceptions of campus racial climate on Latino college students' sense of belonging. Sociology of Education, V 70 (4), p 324–345.

Independent Commission on Fees Report, (2014) The Sutton Trust: Improving Social Mobility for 20 years. Available online at https://www.suttontrust.com/research-paper/independent-commission-on-fees-report/

Johnson, J. (2017) Delivering value for money for students and tax payers, part of Access to Higher Education and Higher Education participation. Available at https://www.gov.uk/government/speeches/jo-johnson-delivering-value-for-money-for-students-and-taxpayers

Langa Rosado, D., & David, M. E. (2006). A massive university or a university for the masses? continuity and change in higher education in Spain and England. *Journal of Education Policy, 21*(3), 343–365.

Lyons, J. (2011). An exploration into factors that impact upon the learning of students from non-traditional backgrounds. *Accounting Education: An international journal, 15*(3), 325–334.

May, V. (2011). Self, belonging and social change. *Sociology, 45*(3), 363–378.

Million+ and National Union of Students. (2012) Never too late to learn: Mature students in higher education Million+/NUS. Available online at http://www.millionplus.ac.uk/documents/reports/Never_Too_Late_To_Learn_-_FINAL_REPORT.pdf. Accessed 7 July 2015.

Nuñez, A.-M. (2009). A critical paradox? Predictors of Latino students' sense of belonging in college. *Journal of Diversity in Higher Education, 2*(1), 48–61.

Office for National Statistics (2017) UK Labour Market: August 2017. Available online at https://www.ons.gov.uk/employmentandlabourmarket/peopleinwork/employmentandemployeetypes/bulletins/uklabourmarket/august2017

Osterman, K. F. (2000). Students' need for belonging in the school community. *Review of Educational Research, 70*(3), 323–367.

Perry, C., and Allard, A. (2003) Making the connections: transition experiences for first-year education students. Journal of Educational Inquiry, 4, 74–89.

Read, B., Archer, L., & Leathwood, C. (2010). Challenging cultures? Student conceptions of 'belonging' and 'isolation' at a Post-1992 University. *Studies in Higher Education, 28*(3), 261–277.

Richardson, S. (2017) Understanding part-time, mature students' 'sense of belonging' when studying higher education in the further education sector. EdD Thesis.

Robbins, L. (1963). *Higher education (the "Robbins report")*. London: HMSO.

Ross, A. (2001a). Higher education and social access: To the Robbins report. In L. Archer, M. Hutchings, & A. Ross (Eds.), *Higher education and social class: Issues of exclusion and inclusion*. London: Routledge Falmer.

Ross, A. (2001b). Access to higher education: Inclusion for the masses. In L. Archer, M. Hutchings, & A. Ross (Eds.), *Higher education and social class: Issues of exclusion and inclusion*. London: Routledge Falmer.

Rowley, J. (2005). Foundation degrees: A risky business? *Quality Assurance in Education, 13*(1), 6–16.

Strayhorn, T. L. (2012). *College students' sense of belonging, a key to educational success for all students*. Abingdon: Routledge.

Strayhorn, T. L. (2013). College students' sense of belonging: A key to educational success for all students. *The Review of Higher Education, 37*(1), 119–122.

Thomas, L. (2002). Student retention in higher education: The role of institutional habitus. *Journal of Education Policy, 17*(4), 423–442.

Thomas, L. (2012) *Building student engagement and belonging in Higher Education at a time of change: final report from the What works? Student retention and success programme*. Available at www.heacademy.ac.uk/resources/detail/retention/What_works_final_report. Accessed 15 July 2016.

Trow, M. (1974). Problems in the transition from elite to mass higher education. In *Policies for higher education* (pp. 51–101). Paris: OECD.

Trow, M. (2005) reflections on the transition from elite to mass to universal access: Forms and Phases of HE inmodern societies since WWII. International Handbook of Higher Education.

Universities, UK. (2013). *The power of part-time, a review of part-time and mature higher education*. London: Universities UK.

Universities UK. (2015a). *Supply and demand for higher level skills*. London: Universities.

Universities UK. (2015b). *The higher education green paper, fulfilling our potential: Teaching excellence, social mobility student choice*. London: Universities.

Warren, D. (2002). Curriculum Design in a Context of widening participation in higher education. *Arts and Humanities in Higher Education, 1*(1), 85–99.

Index

© Springer International Publishing AG, part of Springer Nature 2018
M. Dastbaz et al. (eds.), *Smart Futures, Challenges of Urbanisation, and Social
Sustainability*, https://doi.org/10.1007/978-3-319-74549-7

Printed by Printforce, the Netherlands